First to Bomb - The World War I Diary of Lt. Howard G. Rath, Bombardier, 96th Aero Squadron

Howard G. Rath
and
Hugh T. Harrington

Hugh T. Harrington
3801 Village View Drive, Apt 1125
Gainesville, GA 30506
hharring@charter.net

Copyright © 2021 Hugh T. Harrington
All rights reserved.
ISBN: 978-0-578-32518-7
Independently published

FIRST TO BOMB

Dedication

To the Red Devils of the 96th Aero Squadron and all American airmen who have fought in the skies.

FIRST TO BOMB

FIRST TO BOMB

Contents

List of Illustrations vii

Preface 1

Chapter 1
Introduction to Lt. Howard G. Rath 3

Chapter 2
The Voyage to War 7

Chapter 3
Bombing Instruction at Clermont-Ferrand 7th Aviation
Instruction Center 15

Chapter 4
To the Front for Bombing Mission Experience 27

Chapter 5
After Training, Awaiting Assignment to Squadron 37

Chapter 6
96th Aero Squadron 43

Chapter 7
The First Bombing Raids 53

Chapter 8
Major Harry M. Brown Flies into Infamy 65

FIRST TO BOMB

Chapter 9
    The Squadron After Major Brown 71

Chapter 10
    The War in the Air Heats Up 79

Chapter 11
    The St. Mihiel Offensive September 12-16, 1918 93

Chapter 12
    Meuse-Argonne Offensive 109

Chapter 13
    Postwar 123

Appendix
    Rath's Letter Describing his first Bombing Raid 127

Acknowledgements 133

Bibliography 135

Index 145

List of Illustrations

Figure 1. The diary 5
Figure 2. 7th A.I.C. Bomb Sight 17
Figure 3. Conflans, with north to the right and south to the left. 19
Figure 4. Rath's piece of flying suit 31
Figure 5. Leaving Clermont for Amanty, May 22, 1918 45
Figure 6. Insignia, guns and observer's window 46
Figure 7. Participants in the first bombing raid plus Royal Flying Corps spectators. Left to right: Newberry, Tucker, Lewis, Beverly, Duke, Capt. Ward (Royal Flying Corps), Maj. Gray (Royal Flying Corps), Strong, Rath, Maj. Brown, Mellen, Browning, MacDonald, Tichener, Smith, Ratterman, H. Thompson, and Evans. 54
Figure 8. Map of operational area for the 96th Aero Squadron .. 56
Figure 9. Breguet 14 B2 81
Figure 10. German photo of crashed unidentified 96th Sq. aircraft. 96
Figure 11. Rath and Bradley J. Gaylord with their DSCs 119

FIRST TO BOMB

Preface

In 2019, during research for *Destiny's Wings, Four Months in Day Bombardment: The Story of Lt. Hugh S. Thompson, 96th Aero Squadron, U.S. Army Air Service in World War I*, I came across World War I historian Gerald C. Thomas, Jr.'s reference to Lt. Howard G. Rath's diary as a "most valuable source of unpublished material." Naturally, I wanted to find a transcription of the diary. I looked in every library and archives I could think of with no result. Not a trace could be found of the original diary, and it had never been transcribed.

My wife, a skilled genealogist, offered to track descendants of Howard G. Rath as maybe they could tell me where the diary was housed. She found a descendant and I wrote a letter. My book was, at that time, going to press.

I was at my desk when the phone rang. I gasped when I saw the incoming name on the phone...Rath. It was Lt. Howard G. Rath's grandson. I asked if he knew where the diary could be found, and he replied that he had it in his hands at that moment. After a short talk, he asked if I would like to borrow it. I will never forget such kindness extended to a stranger.

In due course the diary appeared. I realized immediately why it had not been transcribed, as it was, in most places, largely illegible. In the spring of 2020 with the Covid pandemic curtailing many activities, I set about transcribing the diary. This involved poring over it with a strong magnifying glass and bright light. To say it was slow going is an understatement.

However, it was not long before I realized that I was deciphering a document of historical significance and one that clearly had not been read before, except perhaps some segments by Gerald C. Thomas, Jr.

I consulted with Michael J. O'Neal, Managing Editor of *Over the Front*, the Journal of the League of World War One Aviation

FIRST TO BOMB

Historians, who enthusiastically encouraged me to continue the transcription. Meaningful diaries are rare and those connected with the bombing squadrons are almost unique. He urged me to publish the diary and suggested the current format of adding historical context to the diary so readers would better understand Rath. The result is this book.

Hugh T. Harrington[1]
Gainesville, Georgia

November 5, 2021

---

[1] Hugh Thompson Harrington, 1950-

Chapter 1

Introduction to Lt. Howard G. Rath

Fate does not often allow a man to be first. Howard G. Rath however would carry, forever, the honor of being the first American, in the first American bombing squadron, to drop an aerial bomb on the enemy.

Howard Grant Rath was not the usual airman. For one thing he was far older than almost all of the men who fought in the air. Born in April 1885, he was 33 years old in 1918, making him about 10 years older than most of those he flew with. He did not have any military background, not even basic or officer training. He graduated from Williams College in 1907 and became a very successful securities broker in Los Angeles. In 1914, sailing west from California, he took an around the world trip, which was cut short when he arrived in Europe via Suez in August just as the war started.[2] Rath returned to Los Angeles while the United States remained neutral, and Europe tore itself apart in the Great War. After almost three years of the world at war, in 1917, President Woodrow Wilson called on Americans to aid the Allies to make the world safe for democracy. Rath voluntarily left his brokerage business to participate in the war effort. He joined the American Field Service sailing from New York on September 22, 1917, arriving via Halifax in England. The voyage was uneventful and Rath immediately went on to France.

---

[2]The outline of Rath's early history is from: Rath, Howard G., *Origin and Development of American Aerial Day Bombing During World War I*. Unpublished typescript, circa 1940's.

FIRST TO BOMB

The Diary of Lt. Howard G. Rath

The Diary of Lt. Howard G. Rath has never been published. The diary is justly considered "one of the most valuable sources of unpublished material."[3] Because the diary was written as the private memoranda of Lt. Rath and not for the public, it carries a much higher degree of candor than one meant for the public eye. The diary consists of over 250 handwritten pocket-sized pages and amounts to over 25,000 words beginning with Rath's voyage across the Atlantic in September 1917, then after a gap of 4 months picks up in February 1918 and continues daily until the end of the war.

The diary is a day-by-day account of one perceptive man's observations, experiences, and thoughts during the crucial period when the United States was leaping into a world war in the air and was quite unprepared for the task. Virtually everything required to become an effective bombing force had to be learned by the airmen for themselves. There was no manual, no experienced advisors to lead the way. The diary tells this story as none other can.

The World War I Diary of Lt. Howard G. Rath was written in two volumes. The diary skips back and forth between the two volumes, but the dates are clearly identified. The dates covered or missing are:
   September 25, 1917 - October 18, 1917 (Volume 1)
   October 19, 1917 – February 11, 1918 (missing)
   February 12, 1918 - March 25, 1918 (Volume 1)
   March 26, 1918 – April 13, 1918 (missing)
   April 14, 1918 - September 16, 1918 (Volume 2)
   September 16, 1918 – November 21, 1918 (Vol. 1)

The Diary is presented here in chronological order.

The editor has made every effort to transcribe the diary exactly as written. Punctuation has been added only when necessary for clarity. Most spellings and abbreviations have been left as Howard Rath

---

[3] Thomas, Gerald C., Jr., *The First Team: Thornton D. Hooper and America's First Bombing Squadrons*, Dallas, Texas, The League of World War I Aviation Historians, 1992, p. 122.

originally wrote them. Clarifications and identifications have been inserted in [brackets]. Rath's handwriting is often very difficult to read. Words that cannot be deciphered are indicated as [*illegible*] or where uncertain with [*?*]. Words in (parenthesis) are those of Howard Rath.

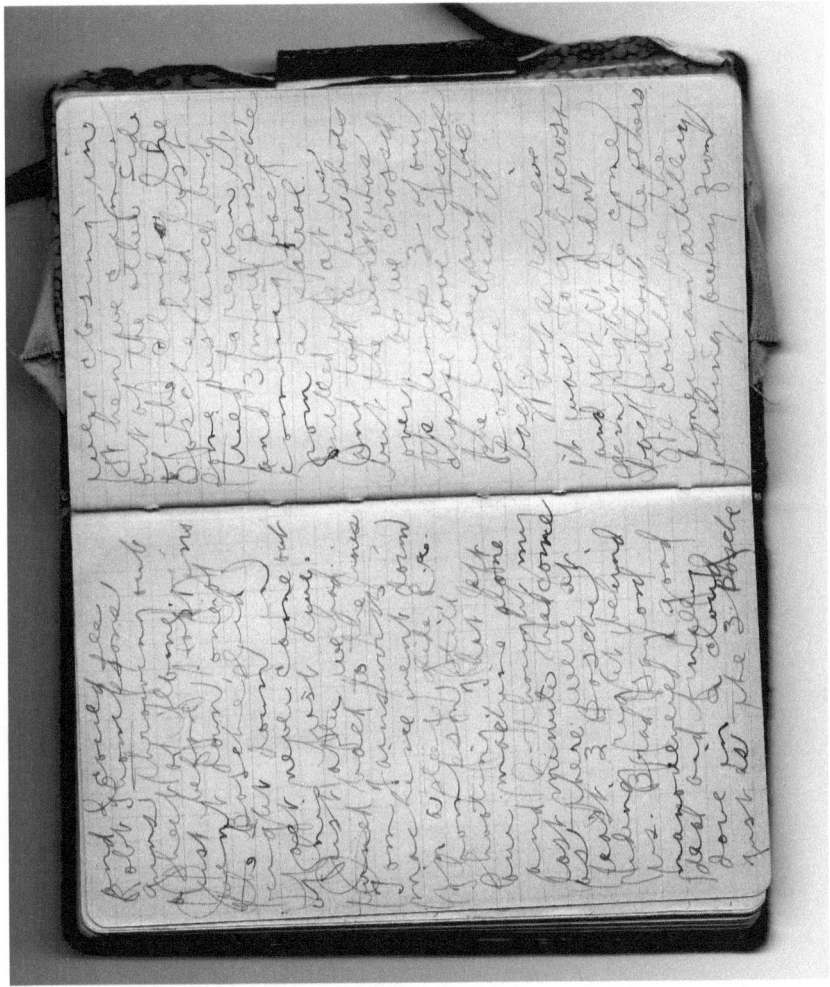

Figure 1. The diary

FIRST TO BOMB

To render the diary more useful, elements of the history surrounding Howard Rath, the 96th Aero Squadron, and the First Day Bombardment Group have been added in *italics* by the editor. I have broken the diary into chapters to make it easier for the reader to locate areas of interest. The original diary does not have chapters.

Chapter 2

The Voyage to War

The diary begins as follows:

Volume One –
    H.G. Rath. August 25, 1917 to [blank]
"The Great Adventure to Somewhere in France" from Somewhere in America

September 25, 1917
    This is the real start of the big adventure. Didn't get to bed until 3 o'clock the night before as it took hours to pack. Had gone to the theatre earlier in the evening with Hun Verity and Florence Kelsey. What a happy coincidence to run across Hun in the big city. She saw me walking thru the Waldorf and that [is] the way we happened to meet. She is going to knit a sweater for me – there is no question about its keeping me warm.
    But to return to this eventful day – after much confusion we got started – said goodbye to the Verity's – piled our baggage in a deep sea going taxi and went down town to get my money changed into a French draft. We were afraid we wouldn't make the boat as the taxi driver was very slow. Were driving down Hudson street when a little girl who was crossing the street suddenly darted right back in front of the taxi. The driver swung around to avoid her but the side of the car hit her. It was terrible to hear her scream. A crowd gathered at once, she was carried to a drug store – and they found that she had a broken

leg and was cut around the mouth. I am sure that it wasn't the driver's fault and we made a statement to that effect. We went down to the boat to put our baggage aboard – then came back and did all we could. Girl was taken to hospital and was getting along all right.

The *Saxonia* was to start at twelve but didn't get away until 4:30. There are 150 aviator cadets on board going to England for training – fine fellows – about 1700 troops mostly infantry and about thirty officers including two major generals and one colonel. Major general Bell is aboard. Every berth is taken – only a few civilians aboard – all English – and only five women. All of our unit were in uniform. The officers can't make us out. Had good dinner – second sitting. Afterwards went on deck – beautiful evening. Guards are posted everywhere. Feel sorry for infantry – they do the hard fighting and get the worst treatment. Corporal tells us about his experiences on the border – evidently used to be a railway brakeman. Says the ship hasn't enough joints [?] in her to make him feel at home. Turned in about 12.

September 26, 1917

Officers objected to American Field Service men being on top deck in the army uniform. Said that it was regular army uniform and they couldn't make exception. We foxed them by going back to citizens clothes. Kept running out of sight of land. Bunny [*John "Bunny" Pike, Assistant Section Director, American Field Service*] gets a little seasick. Soldier boy waits on our table – a tip makes everything come our way. We are on our way to Halifax as a rendevous [*sic*] for convoy. Some of the fellows get a little liquored up. Elliott [*Walter Keith Elliott, AFS*] the Boer War "veteran" is a card. He says nobody can give him orders as he is a major general in his own estimation. Taylor [*James J. Taylor, AFS*] of Hollywood tells us all the movie scandal. Says that Mary Pickford and Douglas Fairbanks have had an affair. Charley Chaplin and Edith Purviance [*Edna Purviance, 1895-1958, actress*]. Weather wonderful – sea as smooth as glass.

September 27, 1917

Another wonderful day – Indian Summer. Great sleep. Sat around on deck and talked. Concert in the evening – very good. Began entering Halifax outer harbor about eight o'clock – anchored about 11 p.m. Bill Reagan [*William N. Reagan, Head Section Director, American Field Service*] sure is a fine fellow. Should be – he is Chi Psi – certainly was a strange coincidence. How soon one becomes accustomed to changed circumstances. Have almost forgotten about submarines our only care now is to get "somewhere in faster time." Had a little close harmony on deck until ship captain said he wanted to sleep. That was sufficient. Wonderful moonlight but of no use? Except to the U.S. Lieut. who is queening the only available girl.

September 28, 1917

Another great day. We were anchored all day in Halifax harbor. There are about a dozen boats. Horse transport – Belgium Relief Boat,[4] Canadian troopship. About noon the *Andania* [*RMS Andania*] began to steam in the harbor. She is simply loaded with American troops – mostly engineers – 1700 about – she passes our ship with all men cheering and both ship bands playing. Didn't realize so many troops were on their way over. All high commanding officers go ashore at noon. One of the generals is rather feeble. The rest of us try to amuse ourselves looking at scenery. This is a beautiful harbor – a regular Evangeline setting – wooded hills leaves just beginning to turn. Had a little session of the great American pastime in the evening – got cleaned.

September 29

A little more wind today but still bright and fair. Everybody conjectures as to when we pull out. Smith [*Leigh Hackley Smith, AFS*] has been running a roulette wheel – is cleaning some of the fellows up. Never saw so many gambling games on a boat. Soldiers surely gamble for high stakes.

Some (few) of the officers get chummy. Especially aviation Lieut. from Davenport, Iowa. Would say that some of the officers aren't any too well versed in their positions. Fellows getting quite well

---

[4]The Commission for Belgian Relief was a primarily American organization that negotiated with British and German authorities for the safe conduct of ships loaded with food for Belgian and French citizens in territory occupied by Germany. The ships carried huge "Belgian Relief" signs painted on their sides in hopes German submarines would allow them to pass. Some were torpedoed.

acquainted. Discover we have some musical talent on board. About 3 o'clock the boats begin to stir a little and at 4:15 in the afternoon we pull the "hook" up and start for the outer harbor. Discover two battleships (English) in the outer harbor and several more passenger boats filled with American troops. What an inspiring and glorious sight it is and how happy it makes one to play his little part in it. As we pass each ship the crew and soldiers line up and give us three cheers while their band plays the "Stars and Strips." We pass the *Lapland* [SS *Lapland*] among other boats and she is just loaded with troops (American). The folks at home would be surprised to know so many are on their way. Boat after boat falls into line until there are nine of us. Among them are the *Andania*, *Baltic* [RMS *Baltic*[5]], *Grampian* [RMS *Grampian*], a tanker, horse ship and several troop ships and a converted cruiser as convoy. We line up in three column of three each after we get outside of the harbor. The *Saxonia* leads the extreme left column (nice position). We proceed with only stern lights – no funnel smoke with exception of boat behind us which leaves trail of black smoke behind her. The adventure is now on in full swing. One of the most beautiful nights I have had at sea. Glorious full moon and smooth sea. The [*sic*] is a sense of comfort to see eight other boats steaming along with our [*sic*]. How silent and unmovable they seem with no smoke or lights.

September 30 – Sunday first rain

A dreary [*sic*] is productive of fears and rumors. From some unapproachable portion of the boat the report emanates that the New York paper has it that the *Saxonia* has been sunk. What an imaginary feeling it gives one to be sleeping dry blanketed on a sunken boat.

The first services are held on the boat and I hear for the first time the combination verse of America and God Save the King. It is one of the best verses to a national that has come to my notice.

"Hester" Newell is still the butt of all the boys jokes. There is such a thing as being a too complacent Jonah – Newell is always waiting for the whales mouth to open.

---

[5] RMS *Baltic*, all but forgotten today, is remembered by those interested in the *Titanic* tragedy as being one of the ships that had radioed an ice warning the evening before the Titanic's sinking in 1912.

FIRST TO BOMB

The inspection – the girl onlooker.

The Irish soldier boy who says he can grab the food while the English steward is thinking

October 1 – Monday – Rainy warm

The first day of a new week and a new month. Would I want to see ahead to the closing day? The ship has settled down to a half drowsy routine. We [*no*] longer take notice of the various changes in the formation of the nine ship. The soldiers drill and relieve guard unnoticed. Passengers are getting bold enough to conjecture as to when we will reach Liverpool. But as darkness comes on old grisly Rumor comes back – this time it is to the effect that a periscope was on our port bow – this must come from the bilge or the barbershop. The barber as all ship barbers shaves beards and grows tales. The best informed travelers' story is bound to get a close shave by his garrulous tongue. What a gleam comes in his eye as he hears some one begin [*to*] unwind a tale and how his jaws stand half open like a pair of shears to cut it down in its full bloom of astounding wonder.

October 2 – Tuesday – Fair

Not an eventful day by any means. The time drags worse than ever for the captain of the ship has decided that there shall be no more playing of the national game. Oh! for a real American meal. This constant diet of boiled carrots, boiled potatoes, boiled beef and anemic white rolls (such as the English alone can like) begins to pall on one. No wonder the English are so conservative – they have "eaten it" all their lifes [*sic*].

"Bunny" Pike has a mania for rooting through all his trunks and bags at least once a day for some insignificant thing – such as a letter. Our stateroom looks like a Kansas farm after a cyclone has hit it. Took great pains to arrange all my clothes in order for a rapid dressing should the emergency arise – even to putting socks in shoes. But its of no use with Bunny around. Couldn't locate my shoes this morning and one of my socks was in Bunny's bed. I've got my life belt hung up on the ceiling – so far he hasn't taken it to bed with him.

Great talk back in Newell's room. Taylor [*James J. Taylor, AFS*], Morse [*Charles F. Morse, Jr., AFS*], Knisely [*George Knisely, AFS*], Pike, Reagan, Lane [*Lewis Palmer Lane, AFS*] & myself. Bunny Pike launches out on the wonders of California & Lane starts an argument on the Japanese question.

October 3-4 Wednesday & Thursday

Two uninteresting days. Dull grey weather & chilly winds. There are now ten boats in the convoy as one caught up with us Tuesday night.

The men in the Amer. F.S. [*American Field Service*] are beginning to figure where they will land as to service. Certainly all of us have seen enough of a private privation not to hanker for it. But you must experience to learn, one has been able to get a better insight of army in this two weeks at sea than in 3 yrs of reading at home. Great gods! These English meals, you know just how the boiled potatoes and roast beef are going to look before you reel into the smelly dining room. And then the English steward who thanks you after you take your helping – and he certainly is doing the proper thing.

Thursday I felt a little squeamish – what do you know about that! was never so dizzy on the water – didn't seem like sea sickness. It's a pretty sight to see the ships maneuver – sometimes we are in the lead – sometimes in the rear.

October 5th Friday – stormy

We are in the submarine zone now. Very strict orders what we can & cannot do. Must wear lifebelts continually. No noise on deck. Had a night boat drill. Ran into lots of phosphorus. Dark nights and this phosphorus should help us. The "subs" will have to go some to get us as the boats are continually maneuvering. One boat does a little target practice. Bunny Pike has a worried time deciding where he will sleep. Wants to sleep on deck but none in our stateroom will go. Finally decides to stay with us but doesn't know how many of his clothes to take off. Asks Wolf [*William B. Woolf, AFS*] if he has his shoes off then wants to know if I have my shoes off. Gas attack. All get to sleep about two o'clock.

October 6th Saturday – foggy – rainy

The first day in the sub zone has been uneventful. The Canadian Capt. got an awful stew on last night and broke up some furniture. Charley the Deck Steward and Drink Purveyor was a close second. He

FIRST TO BOMB

asks us if we want some "Binger Jeers" and then he gets lost in the smoking room bringing us our Ginger Ale. Our crowd has now reached the point where "like associates with like." There are three or four well established cliques. Smith [*Leigh Hackley Smith, AFS*], Elliott [*Walter Keith Elliott, AFS*] et al., LaForge [*Edward C. LaForge, AFS*] et al., Pike, Reagan, Rath et al. etc. "Peter" Newell [*Hester"Peter" Newell, AFS*] still raves about his girl. Looks at her picture at the table at every meal.

October 7th Sunday – Windy

Picked up our convoy this morning. Eight destroyers – four American, four British. American are camelflougd (?). British are not. British are built lower and do not show so much. Went to church on after deck. Poor services – struck me as if chaplain was anxious to get through with them. Slept in afternoon as didn't get much sleep at night. Bunny has a fussy time getting to bed, has to find out if Wolf is wearing his trousers in bed, wants to know if I got my shoes on. Finally he goes to bed with his clothes on. Big discussion in evening about Field Service. Everybody in doubt about it. Bill Reagan finally shows he has the authority and things straighten out. Rough weather is following – all to our favor. A sub would have to act quick to get us in this sea.

October 8, Monday – stormy

The worst night yet. Went to bed undressed. Was awakened at 4 o'clock by crashes of breaking glass & tumbling furniture. Some one yelled "all hands on deck" and what a scramble there was. Wolf piled out from the bunk under me and was dressed in a jiffy. Anderson said "she's hit fellows" and he began to dress. Bunny who was all dressed was trying to put his overcoat over his trousers. Got up on deck to find it all a false alarm. General Wright was up in smoking room with trousers – lifebelt & unlaced shoes. Bunny Pike and some of the fellows try sleeping on deck. Wolf and I go back to bed and have a good sleep. Later discover Bunny back in bed with his clothes on & life belt tied around him.

[There is a gap in the diary entries of four months – blank pages – until February 12, 1918 when the diary picks up again]

FIRST TO BOMB

Chapter 3

Bombing Instruction at Clermont-Ferrand 7th Aviation Instruction Center

In Paris, Rath experienced some unspecified difficulties with the ambulance service and went to work, as a civilian clerk, on October 19th at the American Aviation Section headquarters. There he quickly learned that he might be able to qualify to take flight training. During the oral examination he discovered that candidates must be no older than 28. He was 32, 5 months shy of his 33rd birthday. Major Edmund L. Gros, MD [1869-1942], chairman of the examination committee, waived the age limit restriction saying that if Rath was willing to get killed for his country he would not stand in his way. Rath passed the oral examination. The following day he passed the physical examination.[6]

While at headquarters Rath became acquainted with Captain Frederick T. Blakeman [1883-1964], who was going to Clermont-Ferrand to help organize the 7th Aviation Instruction Center (7th A.I.C.), which would be the American bombing school. He requested that Rath go with him. Blakeman and Rath, who received a direct 2nd Lieutenant commission on December 16th, arrived at Clermont-

---

[6] Rath, Howard G., *Origin and Development of American Aerial Day Bombing During World War I,* unpublished typescript, undated circa early 1940's, p. 2-4. Much of Rath's enlistment and training information described here comes from this source.

FIRST TO BOMB

Ferrand on November 17th.[7] On November 21, 1917 Rath was officially assigned to the 7th Aviation Instruction Center for training as an Aerial Day Bombing Observer, a "bombardier" in modern usage.

Rath and Lt. Pennington Way [1890-1918] were the first students to take the initial bombing instruction. Rath and Way completed their training in mid-January. They were immediately made instructors. The first class of actual teams of pilot and observer bombing students began on February 20, 1918. The pilots and observers paired off into pilot/observer teams of their own choosing. They chose each other based on compatibility, as their effectiveness in bombing and their ability to communicate over the engine, wind, and gunfire noise was essential. They would bond together to the point that each knew almost instinctively what the other needed and when it was needed. On the ground the teams often ate, slept and spent their free time together. After training the teams would be sent to the front together.

Formation flying was stressed for pilots as well as simulated raids, attacks by pursuit planes, navigation, cross-country flying and gunnery.

Much of the instruction for the observers was in navigation, map reading, gunnery and practical use of the all-important bombsight. The Rolling Carpet, an ingenious training device consisted of a wooden box positioned above a realistic terrain map fixed onto a moving treadmill. The view from inside the box allowed the student to learn to use the bombsight as he viewed the passing landscape below. It could be considered a primitive simulator.

Another device was the camera obscura, which was a darkened tent or room where the image of an aircraft could be projected onto a table by way of a lens. The student bombardier/observer could follow the movements of the image and calculate the aircraft's speed and course.

The bombsight was designed and built by the Michelin Company. Within months the students and staff at the 7th A.I.C. would help assist the Michelin designers with modifications resulting in a new sight

---

[7] Gorrell, Series J, vol. 7, p. 211. For Rath's commission see Rath's Officer Record Book.

known as the 7th A.I.C. Sight, which proved to be of high quality and would become standard.[8]

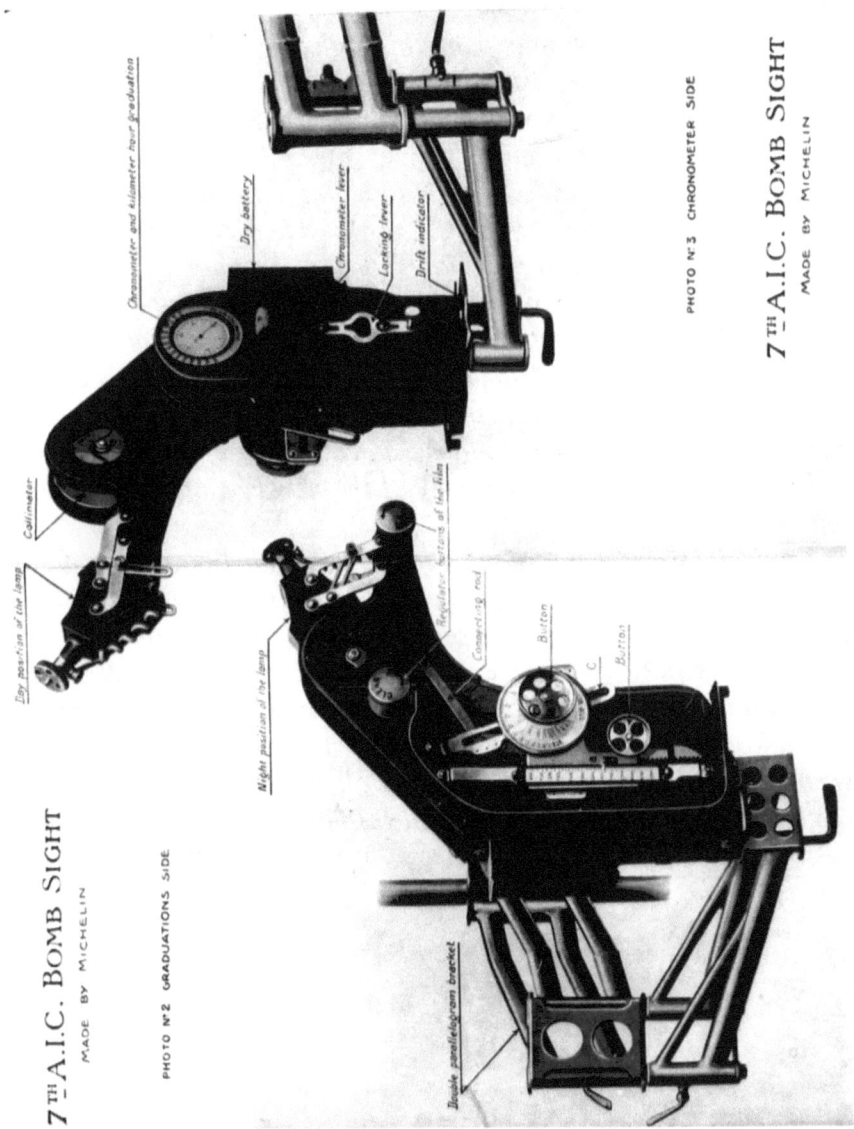

Figure 2. 7th A.I.C. Bomb Sight.

---

[8] The instruction manual for the *7th A.I.C. Bomb Sight, made by Michelin & Co., 1919* can be found in Gorrell, Series J, vol. 4, Report on Day Bombardment Training in A.E.F., https://www.fold3.com/image/19197050.

Rath, and others during the war, referred to their bombing as "precision." The 7th A.I.C. Bomb Sight manual optimistically states that the bombsight allows "the bombarder [observer/bombardier] to drop the bombs with precision." While the sight was sophisticated for its time, the bombing results from this sight, and all others, was not precision. Accurate precision bombing, even during WWII with the famed Norden bombsight, was unobtainable until the advent of smart bombs in the late 20th century.

The bombsight was installed within the observer's rear compartment rather than hanging over the side of the fuselage as did other models. The bombsight, and the observer, could look through a slot in the floor far in advance of the aircraft to observe the landscape for navigation and also establish visual contact with the target.

The bombsight could take into consideration wind speed and direction, air speed, elevation, and drag coefficient of the bombs being used. However, determining those essential variables exactly was nearly impossible making accuracy at the usual bombing altitude of 10,000 feet very unlikely.

Railroad yards were frequent targets not only because damaging them would interrupt enemy transport but also because they were large so they could not only be seen from the air but also more likely to be successfully attacked. A favorite target of the 96th Aero Squadron to which Rath would eventually belong, was the railroad yard at Conflans; it was so big it could be attacked from any direction, and precision was not absolutely essential.[9] The yard at Conflans was roughly 250 yards north to south and 1100 yards east to west, making it a significant geographical feature.[10] Yet, it was often missed as will be seen.

---

[9] Leiser, "Red Devil in a Breguet, David H. Young, 96th Aero Squadron," p. 160.

[10] Harrington, Hugh T., *Destiny's Wings, Four Months in Day Bombardment: The Story of Lt. Hugh S. Thompson, 96th Aero Squadron, U.S. Army Air Service in World War I*, Gainesville, Georgia: privately printed, 2019, p. 80.

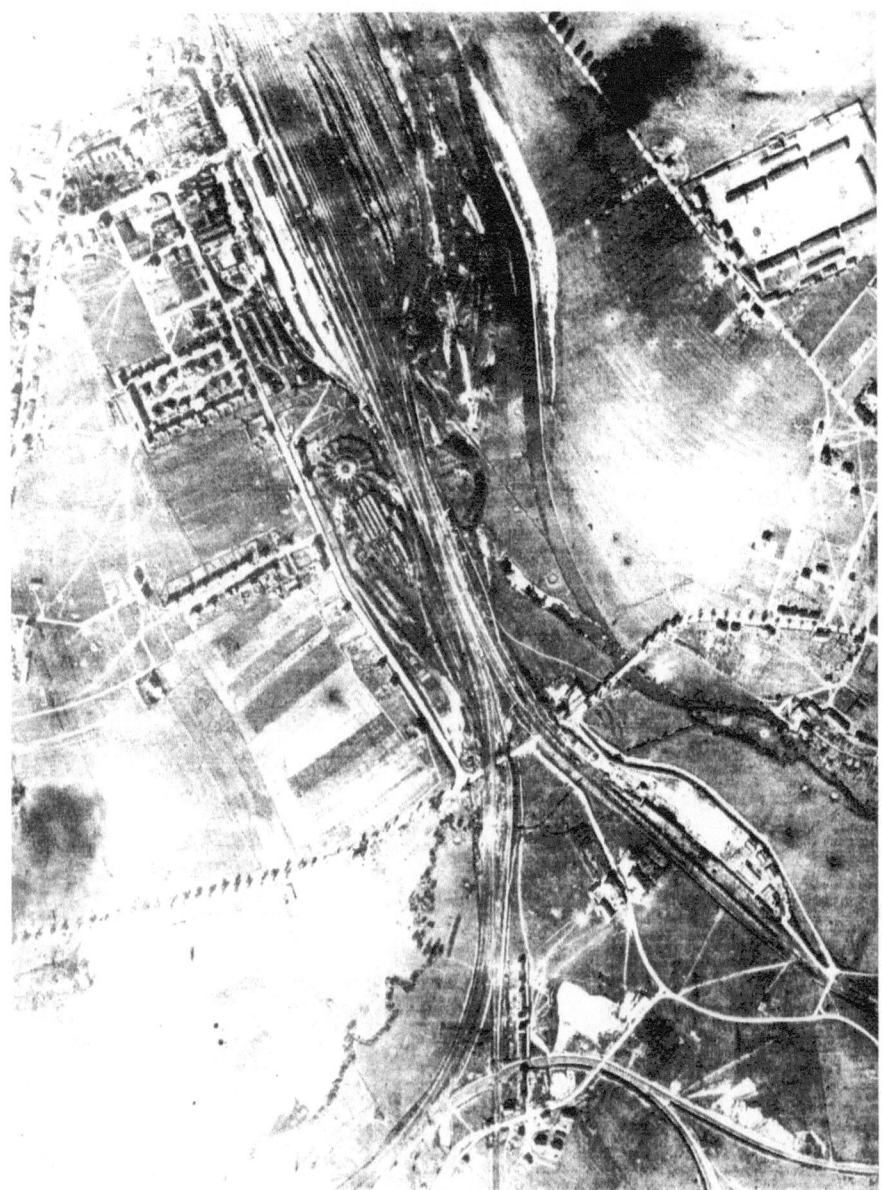

Figure 3. Conflans, with north to the right and south to the left

FIRST TO BOMB

Feb 12, 1918

Was over at C.F. [*Clermont-Ferrand, site of the 7th Aviation Instruction Center*] last evening. Up early this morning. Looks like a good day for flying. My name wasn't on board this morning. Lost out on bombing formation this morning. Got up on machine gun work in afternoon. Not very good – dusted target once. Went on bombing trip afterwards Red [*Pennington H. Way, also known as "Penn," "Pinky" or "Red"*] lead - not very good.

Feb 13

Machine gun work this a.m. Dusted target. Lead bombing formation afterwards Rotten way off. Girl YMCA arrives in camp. Seems good to see American girl. Machine gun work in afternoon. Don't think I did much. Hanby [*Haley?*] had a hard time landing. Finally got down but broke both wings. I got all braced for a tip over. Girl sits at our table – She is from Wellesley and a school teacher. Started a letter to the University Club [*University Club of Los Angeles*] this eve.

Feb 14

Rain today – Did machine gun work this a.m. Went to town in afternoon. Ran across Capt. Reel's friend – professor. Found out only couple teams go out at a time.

The Americans at Clermont-Ferrand were unable to find machine guns in the American 30-06 caliber. However, they did acquire Vickers and Lewis guns chambered in .303 British. This was the standard ammunition for a British infantry rifle. The machine guns were standard infantry issue and were modified for use in the air.

The forward firing Vickers gun was mounted on the left side of the fuselage and operated by the pilot. The water-cooling apparatus was removed, and slits cut in the guns radiator allowing the fast moving cold air at high altitude to enter and provide cooling. Being immobile the gun was aimed by pointing the aircraft itself.

The standard Vickers gun would fire 450 rounds per minute. Due to the nature of aerial combat, where targets might be in the sights for only an instant, an increased rate of fire was desirable. The Vickers was modified to fire up to 800 rounds per minute by adding a muzzle brake onto the muzzle to provide more recoil to operate the firing mechanism faster.

The Vickers was fired by the pilot through the use of a flexible wire known as the Bowden control that was attached to a trigger on the control stick. A complex mechanism operating off the engine crankshaft interrupted the firing mechanism of the machine gun, allowing it to fire bullets past the spinning propeller blades.

The machine gun on the side of the fuselage was not directly below the gunsight that was in front of the pilot's face. The sight and the machine gun were harmonized so that at a pre-determined distance ahead the bullet and the view from the sight would converge together on the target.

Rath, being the observer, would sit in the rear seat position where he would operate the bombsight and also twin Lewis machine guns. When firing the guns he would stand to give himself better mobility, view and broader field of fire.

Initially, the Lewis guns were of the infantry type and were modified in a similar manner as the Vickers gun. A few months later the guns supplied were of the aircraft version. The two guns were mounted on a yoke called a jumellage, which was then attached to the scarf ring (tourelle) that consisted of two rings, one moveable and the other mounted, so it encircled the observer's cockpit. The lower ring was attached to the fuselage. The upper ring turned on the lower ring. The rotary action of the tourelle, combined with a joint, permitted the perpendicular action of the jumellage. This configuration made it possible to aim the guns in all directions, except directly rearward below the horizontal stabilizer or forward so the aircraft could not be hit by its own machine-gun fire.

Both guns were fired together using the Bowden control. The muzzle brake recoil enhancer increased the rate of fire from 500 to 650 rounds per minute. The Lewis guns were fed by flat, circular magazines rather than belts as used in the pilot's Vickers gun. The Breguets used the 97-round flat magazines with five spare magazines carried in a rack in the observer's compartment.[11]

---

[11] *Handbook of Ordnance Data, November 15, 1918*, Washington, Government Printing Office, 1919, p. 249-254.

FIRST TO BOMB

The observer would wear a heavy leather belt around his waist. Two straps connected the belt with the tourelle, allowing the observer to stand and turn with the tourelle. It also served as a safety belt during maneuvers of the plane.

Feb 15

Got in a few shots with the machine gun. Weather went bad and had to knock off work.

Feb 16 – Saturday

The trial of Valentine is coming off Monday. Capt. Thomas [*Capt. George C. Thomas*] has turned against the defense. Looks rather dubious for Val.

Weather turned cold this morning. Did machine gun work and for once at least Gundelach [*Lt. Andre H. Gundelach, 1895-1918*] admitted I did good work. Has promoted me to senior wing of gunnery. In afternoon went on a formation flight – went up with Greene – very near froze to death. Read in paper that Fitzgerald & Bob Graf got the Croix de Guerre in G.B. 13 [*Groupe de Bombardment (i.e., Bombing Group) 13*]. Went to town in evening. Talked to people in Salle de Epichy [*?*]. They also say I look like Judey. [*Howard Rath's nickname was "Judy" or "Judey", perhaps a reference to Judy of Punch & Judy*]

Feb 17 – Sunday

No work today. Slept late in morning. Told Metzger [*Lt. Arthur R. Metzger*] about my trip around the world. In afternoon major, Dr Blackshear and I took a walk – about 8 miles – Major got a little tired. Capt Thomas working hard on case of tomorrow. Looks rather black for Valentine. Tried to sew on wing but made a bad job. Talked about case of tomorrow. Some student pts [*pilots ?*] arrived.

Feb 18 – Monday

Doctors all worked up about Valentine's trial. Capt. Thomas thinks he has a good case against Valentine. Looks like some witnesses have a lapse of memory. Capt. Kelly [*?*] is back for trial. Near noon case takes a decided turn for defendant. Capt Boyd can't keep up to Garrett's speed. Garrett makes a great speech in closing.

FIRST TO BOMB

Everybody decides the defendant gets off. The colonel "Rip" presides. Capt Reel comes up in room afterwards to celebrate. Didn't get to bed until 12 o'clock. Rest of student Lts arrive. Put wing on for first time.[12]

Feb 19 Tuesday

Everybody seems to feel good about trial. Making changes in headquarters. Hear that some must move. Wonder if I go. Discover it is to be Roberts, Lee & Woolley. Cadets come in on evening train. Discover today that Way [*Pennington H. Way*] and I are on Instruction Staff. In afternoon Way and I go to town. Run across Metzger rent cello for 4 francs a month. Go in Globe Cafe and the old crowd is there. Very near freeze coming out on the machine. Orchestra in officers club in evening.

Feb 20 Wednesday

School for student officers starts today. Way & I won't come in for work for several days. There isn't much to do in afternoon. Evening sit around and listen to orchestra. Capt. Reel, Capt. Thomas [*Capt. George C. Thomas*] and Lt. Sellers[13] [*Cecil G. "Swede" Sellers*] come in in evening. All sit around and crack nuts on floor. Major who sleeps right under us had a hard time sleeping. Nobody left until 12 o'clock.

Feb 21 Thursday

Rain this morning. Good news – our orders come through. Lt Greene, Brainard [*Spencer Brainard*], Robertson [*Philip Robertson, believed by the editor to sometimes be called "Bobby"*]. Way & myself leave Sunday for French squadron at Cernot [?] – leave Sunday at last we get some action. Sew suit in afternoon, afterwards Major comes up and Dr. Blackshear and myself go walking with him.

In evening go in Hammond & Nortons room and listen to concert. Garrett puts some good ones over.

---

[12] The wing is the Observer's single wing insignia that is sewn onto the uniform's left breast.

[13] Cecil "Swede" Sellers would lose his life January 11, 1938 in the crash of Pan American Airways Sikorsky S-42B, "Samoan Clipper," at Pago Pago, American Samoa. He was 1st Officer on that flight.

FIRST TO BOMB

Feb 22 – Friday

The day to celebrate. This is the day that the Kaiser was supposed to start his offensive. Maybe he is waiting for me to come up. Went to town this morning had official photograph taken also bought French kepis. 24 bags of mail today but none for me. Way and I took walk in afternoon and here I am writing this this evening, goodnight.

Feb 23, Saturday

Worked all morning getting ready for trip. No mail. Capt Blakeman [*Frederick T. Blakeman*] not very enthusiastic about commission.

Did aerial machine gun work in afternoon Gundelach put us through paces. Tore my suit again. Got news that we are to go tomorrow morning. In evening had a talk with Miss Gladys MacArthur [*of the YMCA*]. Sewed suit.

February 24 Sunday

Left this morning with Lt. Green [*probably Robly Green*], Lt. Brainard, Lt Way supposedly for the front. Wired Bunny [*John "Bunny" Pike, AFS*] I was going through Paris. Uninteresting trip – rest of fellows played cards most of the time. Got to Paris six o'clock. Howard [*unidentified*] was there to meet me said Bunny was out of town. We were also met by an orderly who says our trip to G.B. 5 [*Groupe de Bombardment (i.e., Bombing Group) 5*] is off. But there may be other arrangements tomorrow. We go to Maxim's for dinner. Afterwards turn in as we are all pretty tired.

Feb 25 – Monday

Bunny wakes me up by pounding on the door. Sure looked good to see him. Have breakfast together. Lt. Greene learns that the trip is off and we are to go back to Clermont in the morning. Isn't that luck.

Do a little shopping in the afternoon. All four of us went to Ville Coubley [*Villacoublay*] [*large French airdrome 8 miles southwest of Paris*] to see the machines out there. Saw the biggest triplane I have ever seen, can't imagine how it flys.

Very cold ride. Went to dinner with Bunny & Bill Reagan [*William N. Reagan, AFS*] at the Marguerite.

FIRST TO BOMB

Joan is gone – [*illegible*] the older girl says. After dinner had a taxi ride – best driver in Paris – a Swiss. Went to bed early.

Feb 26 – Tuesday
Got up early to get train. Chambermaid helps us get ready. Lt. Buckler is on the same train. I read and the other fellows play cards. At St. Germain see an American Red Cross girl – looks good. As we get nearer Clermont the weather gets better. Home seems to be any place that you get to feel comfortable. Go to bed early.

Feb 27 – Wednesday
No work today to speak of. Get things straightened out and take a walk.

Feb 28 – Thursday
Heavy wind today – no flying. Get a little mail. Great concert at ymca in evening by French actors – opera. Great stuff.

Mar 1 Friday
Go to work on the shooting field – monotonous work but must be done. Not very good for flying. Some boxing at ymca in evening.

Mar 2 Saturday
Snowed quite a bit last night. Go out on shooting field again. Good work. Take flying suit out and keep pretty warm. Almost get hit by a magazine that is dropped [*accidentally from an airplane*]. Get lots of letters from home - feel better.
Beautiful sunset this day – yellow tints. Penn [*Pennington Way*], Bobby, Norton & I go into movie – get to laughing.

Mar 3 Sunday
Get a late start. Take walk with ymca Miss Gladys McArthur. Great girl. Sit around in afternoon and talk to Way & Gundelach [*Andre H. Gundelach*]. Try to get bath in city but full house. More letters – great. Beaucoup snow.

[*2 blank pages in diary – next entry is Friday March 8, 1918*]

FIRST TO BOMB

In the first week of March 1918 Russia signed a peace treaty with Germany. This was extremely significant as it allowed the Germans to transfer huge numbers of troops and equipment from the Eastern Front to the Western Front facing the Allies. The Germans knew that their best strategy, perhaps their only strategy, would be to knock the British, and, if possible, the French out of the war before the massive American buildup strengthened the Allies to the point where the Germans would face inevitable defeat.

Chapter 4

To the Front for Bombing Mission Experience

Mar 8 Friday

Leave C.F. [*Clermont-Ferrand*] for Paris. Are on our way again. Brainard, Green, Robertson, Way & myself are to go to G.B. 5 [*Groupe de Bombardment (i.e., Bombing Group) 5*]. Dr. McNeil on train also several girls. Bobby examines engines as we go along. Get to Paris about six. Hotel Continental. Lose Bobby. About 8:30 warning that Gothas are raiding Paris. Lights turned out. Go out in street. See flare and explosion. See machines – turn lights on in sky. See something fall in fire. Hear and see flashes of machine gun on plane. See red lights on wing of plane (French) Archies [*anti aircraft*] are bombing all the time. People rush for taxis – yell at auto with headlights burning. Hear machine in air and archies at eleven every where. I go to bed. Tomorrow we finish our train trip.

Mar 9 Saturday

Do a little shopping in morning. Take train for Chalons [*Chalons Sur Marne*] at 11 o'clock a.m. We all notice a better grade of officers and men on train. The best seem to be at the front. Reach Chalon about 3. No one to meet us. See Chinese coolies at work for first time. Also French soldiers with automatic rifles. Green & Way go to look up some one who knows where we are to go. I look up the American Red Cross and get some chocolate. Was a year for Green and Way to get back. Way says that camion [*truck*] is coming from G.B. 5 for us and Green says he is going to get in touch with G.B. 6.

FIRST TO BOMB

Robertson, Way & myself sit around hotel. Green & Brainard go to get in touch with French. Camion from G.B. 5 arrives but as Green & Brainard aren't around we get dinner at Chalons. While eating Green comes in and springs the idea – that Way, Robertson & myself go to G.B. 5 and that Brainard and he go to G.B. 6. So we part and bump over the roads in the [*sentence not completed*] G.B. 5 Commandant [*probably Commandant Vuillemin or Capt. Petit*] doesn't seem very glad to see us – asks us what we have come up for. Bobby tells him in his most educated English. They make arrangements to billet us out in the town and we follow the man along the muddy street which is only lit up by his flashlight. I draw a pretty good billet and Way says he will have to stay with me because he can't speak any French.

We both climb into the same bed between feather beds – but neither of us sleeps much. We are both bothered by "cooties." They fairly drive us wild.

As part of his training Rath attended the School of Aerial Gunnery, in Cazaux, France. The techniques, and skills required to fire from one moving aircraft at a target not only moving but approaching or moving away at various angles were difficult to acquire. However, it was a vital skill that had to be mastered.

Rath, along with his friend Pennington H. Way, would become the first two students to take the bombing course. Upon graduation both Pennington Way and Rath were made instructors at the 7th A.I.C.[14]

As instructors they were required to participate in at least one bombing mission across the lines to give themselves experience and their students confidence. They flew with the French Groupes de Bombardement 5 (bombing squadron number 5) across the lines into combat. This raid was flown by 22 Breguet 14 B2 aircraft from the

---

[14] University Club Bulletin, Vol. 2, No. 1, April 1, 1918, p. 3. *Williams College in the World War*, published by the President and Trustees of Williams College, 1926, p. 221-3. Rath, *Origin and Development of American Aerial Day Bombing During World War I*, p. 3-4.

FIRST TO BOMB

base at Mairy-de-sur-Marne to bomb the railroad at Amagne-Lucquy 40 miles to the north.[15]

Mar 10 Sunday

Breakfast is at ten today (summer time). Kinsolving [*possibly Charles Kinsolving*] was on the road down to see us and afterwards run into "Mary" Kyle [*possibly George Kyle*]. Sure was good to see them. All the Americans are 1st Lieuts. All the French pile in to breakfast and the Captain calmly announces to Way and myself that we are to go on a bombardment right after the meal. I got so excited can hardly eat. I am to go with a French Lieut who can speak a little English. Penn Way is to fly with another squadron with an American boy by name of [*blank*]. Ride out to the field in an old army camion with a crowd of French flyers. As usual they all sing as they ride along. Go across the canal and the Marne and up to the hangars on the hill. Enormous field and our hangar is way across the field. Some job to pack the bag way across there. All the machines are out and turning over [*the motors are running*] and it looks like business. Take a good look at the map to get the directions and then watch the mechanic get the machine guns ready. Have a wind sight and the two machine guns are linked together. All start off in formation and begin climbing as soon as we get off of the ground. Get a good look at Rheims and first view of trenches. Get to our meeting place in about 50 minutes of flight, reach our altitude of 5000 metres – 2 ½ miles. The sun was shining and I got a fine view of the trenches as we passed the lines. [*They crossed the lines near Rheims.*] The front look just like a great big line of yellow clay. The ground was all chewed up with shells. The greater part of the trenches were on the French side, on the German side they seemed very narrow. Am told this is because the French have captured a good deal of the German trenches. The formation holds good and on the left and in the rear are the other two formations. Hit up a little faster pace when we get past the lines and after we get past about 5 miles the anti aircraft commence. At first I thought that they were German machines that had come up after us but

---

[15] *Suddaby Western Front Bombing Database*, Steven Suddaby, www.overthefront.com, accessed August 1, 2020. See also, Martel, Rene; trans. Allen Suddaby, ed., Steven Suddaby, *French Strategic and Tactical Bombardment Forces of World War I*, (Lanham, Maryland: Scarecrow Press, 2007), p. 319.

soon I see them break. They seem to puff out of a clear sky just like popcorn only they are pitch black. After they appear they look like a flock of black ducks for they are all on the same level. As they break nearer I can hear the wind rush through the wings – wgooh! wgoof! and the plane rocks up and down and side ways. We travel straight on, see a German plane way below us but we pay no attention to it.

Go to our objective which is about 30 minutes past the line and when the French Lieut. pounds on the side of the fuselage I pull the trigger and release the bombs. Can see them drop for a long ways but don't see them explode. We were traveling right along with the railroad track so some of us must have hit it. Turn back immediately and just about that time see four German planes following us. They trail us right along but rather far in the rear. Lt. Way who is in the rear squadron gets in a little brush with them and empties his machine gun at them. No results. The archies [*anti aircraft*] keep popping away at us but we travel right along. Feel easier when we cross the line again and as we turn towards the field little white clouds go scurrying along a mile or so below us. Just as we are going down on the field our motor stops but the Lt. makes a pretty landing. Just as we are coming down I discover that there is a cake of ice on my nose and that my nose and cheeks are badly frozen. Get out of the machine put my things away and the Lieut takes me over to the doctor who dresses my nose. Miss the camion so have to walk all the way back to camp. Nose feels like a big bunch of lead on my face. Kyle comes down to see me and we buy some cookies and take a walk along the canal of the Marne. Kyle tells me he would like to have me come up as his observer. Wants me to see Lt. Blakeman about it. When we go in to supper am surprised to see Green and Brainard eating at the table. They look rather sheepish as they tell about there being nothing for them at the other squadrille. After dinner the Captain tells me I am to go up in the morning again. Way and his pilot had a forced landing way over east so didn't show up. Go down to George Kyle's room and talk with him and Robly Green comes in, was going to sleep with me but finally takes the bed (Clapp's [*Lt. Roger Clapp*]) in Kyle's room. Go to bed about ten but can't sleep well – cooties bother me, also my nose.

This first mission must have meant a great deal to Howard Rath as inside the cover of the diary is a piece of green material 3x4 inches on which is written *"piece of flying suit March 10, 1917* [sic – it was 1918] *Judy Rath First bombardment Over German lines."*

Figure 4. Rath's piece of flying suit

For another account of his first bombing mission, and the days immediately after it, please see the Appendix, which contains a copy of the letter he wrote to his brother on March 27.

March 11 Monday

Get up at six o'clock. Heavy bombardment at the front. Shave and walk down to breakfast. Eat in kitchen. Green and Brainard come in later. Haven't (G&B) been assigned to machines. Ride out to field in camion – the bumps are fierce – all the Frenchmen groan every time we go over a bump.

Find I am to go out with another 1st Lt. Frenchmen this morning. He asks very solicitously if I can shoot the machine gun. I ask for a sighter [*bombsight*] and finally they put an A.J.A.[?] on. Not much good but better than none. Machine are all out in place and turning over but the observation machine comes down and says it is still too foggy. We hang around until ten o'clock when we start out. Engine turns over fine and we keep right up in formation up to our meeting place. The line this morning is covered over with artillery smoke. A big engagement seems to be on. Just as we get up to the line my pilot turns back shaking his head as if the engine was not working right. I keep watching the back end for Bosche but keep one eye on the compass. Discover that the pilot is going straight south – he should go south east. After travelling that way for half an hour I lean over and tell him that his is going too far south and that he should go east. But he shakes his head as if he didn't understand me. Keeps going south until he reaches river with a canal which he mistakes for the Marne. I still insist that he should go east but he turns back south west. By this time he has lost considerable altitude and is down to about 12000 feet. Have been traveling for about 2 hours since we started. He picks out a field where there are three hangars and begins to swoop down. Doesn't tour the field first but slips right down. Just as he is going to settle down on the ground he discovers a road and has to raise over it. There isn't much time left to do anything. He is too low to get over the trees in front of him and he is too high to settle down easy so he just seems to drop her right down.

We hit the ground and bounce about 20 ft up in the air. I thought my time had come as I could see a conglomeration of trees, hangars and French soldiers and I wondered how the mess would look when we hit it. Came down with a crash and stopped with sort of a half turn slide. The old machine stayed together better than I thought it would – soldiers crowded around – seemed to be thousands of them that sprang up from the ground. Get out and see that the machine isn't damaged much except that wheel is blown out. The French Lt. is very excited and talks to every body at the same time. Asks if I noticed the fog – that his engine wasn't working well and so on. Have a guard around plane and we go up to report to commanding officer. Find him in the farm house eating lunch with an officer doctor and another Lt. He himself is a Captain. He invites us to dinner or lunch and we are glad to stay. After lunch we walk to Romilly to telephone – it is some walk – we are quite a curiosity to the people especially me. Can't reach the camp so leave the message and get a shave. By this time my nose is a sight has turned black. The Lt. insists that we walk back to the camp which we do. Get lift part of the way in a camion. Take the water out of the machine and have it pulled into the hangar. Wasn't exactly sure just what the bombs on it would do but feared to take a chance. Just as we are working on it a Spad drops into the field. They are lost also on their way to the front from the factory. They are out of essence so go in the hangar also. Afterwards the Lt. and I walk to town again – buy a little bundle of supplies such as soap, toothbrush, sponge (the French Lt must have his cologne also) and go to the hotel. Hotel de Chenein de Fen [?]. Have good dinner - I pay for it – the Lt has his champagne and gets quite talkative. He takes quite an interest in the girl that waits on us but she is able to take care of herself. They give us candles to find the way up stairs and we again have the feather beds.

March 12 – Tuesday

Wake the Lt up. Get breakfast and walk out to camp. Nobody has arrived to help us yet. We sit around all morning. Beautiful day – Skylarks are climbing up in the sky and singing, peasants are working in the fields. I find a lumber pile and almost go to sleep on it. At noon go into little town for lunch. We have run out of cigarettes and can't buy any. The women [sic] at the cafe is afraid that her dishes aren't good enough for the American. Walk back to the camp but no machine has arrived. Very soon however one begins to come down and for an instant we think it is a Breguet but it is only a Sampson [Salmson] that is lost. The pilot gives the officers (several of them) of the Chasseurs a ride and then leaves. About 3 o'clock another machine comes in but again disappointment for it has come over to look the field over. He is an American captain riding as observer, ground construction officer, who came to inspect the field for an American aviation field. They soon leave and we after buying some cigarettes from the French canteen walk to town. We go another road this time and the French Lt. tries to sing American songs. He keeps singing "John Brown's Body Lies a Moulding 'on' the grave." He thinks it is a song about a drunken man. I explain it to him and also teach him the verse about "We will hang Jeff Davis to a sour apple tree."

Eat a very quiet dinner but get into a big argument about the war. He seems to have gone stale on it. We get switched to religion and he makes the statement that "There must be some sort of a reward for those who make the big sacrifice that may come to one in aviation."

FIRST TO BOMB

March 13 – Wednesday

Get up at eight – breakfast and go to a coiffeur for a shave. Start to walk out to the camp but get a lift in a French camion. Talk to the officers of the Chasseurs until about noon. They take me over and demonstrate with some hand grenades.

They invite us to lunch but just as we are going in we hear a plane buzzing overhead and sure enough it is a Breguet coming down. We go back to the field just as it alights. They have come for us and the French sergeant pilot has a mechanician along. Has a spare wheel for us. We all go into dinner and sure have a wonder. The way the French drink wine is a caution. After lunch put the new wheel on the machine and trundle her out of the hangar.

The French sergeant takes several of the French officers up for a joy ride. Afterwards (about 2 o'clock) both machines are cranked up Lt. de Bourjolly [?] of the Chasseurs gives me some grenades for souvenirs and we get ready to start.

I get in with the French sergeant so that the mechanician can ride with the Lt. who claims his engine isn't working right.

I settle down in the seat thinking my worries are over but as I keep watching the compass I see that we go east instead of northeast. After traveling that way for twenty minutes the pilot turns south. I know that is the wrong way even though we were traveling way to the south of my map. I get up and tell him east and northeast but he won't turn any farther than east. We hit the Marne finally but he insists upon going down south instead of up north. He finally asks me where we are and where Chalons is but I don't know just where we are. He decides to come down and find out. So we circle a wheat field several times which is situated near a small village. Makes a perfect landing – he gets out to ask a farmer where we are. The way I wanted him to go was the right way. He gets in tries to get up in the air but we are out of gas. Peasants come crowding around and the sergeant says that he will go to San Sermaize [? *Possibly Sermaise*] to telephone while I watch the machine. It takes him about 2 hours to come back. While he is gone two American Red Cross boys come up to talk. Also see a plane flying over Revigny. They say it is bombarded every night. The sergeant gets back just as I was going into the village. Says a camion will come out with gas tomorrow. We take the water out of the radiator and walk to San Sermaize [?]. Germans have burned all the houses here in their great advance of 1914. Pretty tired when we get to San Sermaize [?]. Get a good dinner and then get billeted in a big Chateau. Beautiful tapestries, large rooms, I almost get lost in my bed room. There is a big door that opens right into the wall on one side. The wall paper and a picture cover it over. Can't sleep well, trains keep going by all night.

March 14 – Thursday

Walk to machine. Truck brings gas. Get some coffee at house built beside house burnt down by Germans. Woman tells about Germans pointing bayonets at her mothers breast and demanding French officers.

Follow Marne home. We are out of luck with French pilots – Green, Penn & I take walk along Marne canal. Stop at little inn – George Kyle & Clapp come in. Robby [Bobby?] "where have you fellows been." Green in cemetery – his salute.

March 15 – Friday

My nose gives me some trouble – Decide to leave this evening. Green & Bobby stay. Get in Paris just after explosion of munition factory.

Stay in Paris four days – See Pike & Reagan – go to Versailles – Bois [?] Boulogne – Blue Pavilion not much of interest.

March 20-25

Get home and find there is nothing to do until we go but can go on permission if we want to. Lt. Blakeman [?] gives the instruction staff a banquet March 25. Had a great time. Pen & I decide to go on permission tomorrow.

25 –

Go to Lyon in morning train. Stay there for 3 days. Go on to Valence for 3 days – pretty place on Loire river. No Americans quite a change – go to St. Peray [?]. Stay a day at Marseilles – lots of English. English colored trooper asks us if we want to go out to "the piece [?]." He can't just make us out. Lots of shipping. Decide to go on to Nice. Train crowded – met four fine English officers from Italy. Say it is very quiet down there. Go to Ruhl [?] Hotel in Nice and get strong properly [?]. Meet some of the fellows. Ride along the Riviera to the Italian border. Meet two fine American nurses. Miss Tom [?] & Miss Wilson an American lady – Lady Strafford talks to us on the train and invited us to tea at Monte Carlo. Meet an English major – see Sarah Bernhardt's son[16] – he is gambling all her money away. Funny American who comes up in Hotel and asks very personal questions.

Travel back on same train with nurses – they are very shocked at conditions over here. Frenchmen on train scramble for seats. French lady complains about hard time she has to get a seat when she pays full fare while military people ride for ¼ fare.

[*The diary stops here and picks up with "continuation from the little black book," the other volume that had run out of pages at September 16, 1918. The continuation will be placed in its chronological position at September 16, 1918.*]

---

[16] Sarah Bernhardt, 1844-1923, French actress. Her son, Maurice Bernhardt, 1864-1928.

Chapter 5

After Training, Awaiting Assignment to Squadron

April 14, 1918

Left Paris 8 - got good seats – Belgian family on train. Girls speak English – Cadets leaving at depot. Hear I'm going to England – Ellis [*Lt. Arthur C. Ellis*] says so. Some one changed the orders. Going with Squadron. Go back in rooms with docs [*?*].

April 15

[*illegible*] today. Roam around camp. Don't know when we will go out.

April 16

Rains. Captain Thomas [*Capt. George C. Thomas, 96 Aero Squadron, Commanding Officer*] tells us how he got things fixed up. Take walk. Looks like next week. Major Brown [*Maj. Harry M. Brown*] arrives.

April 17

Meet Major Brown – young and has pep [*Brown was 27; Capt. George C. Thomas was 44; Rath would turn 33 the following day*]. He says he is going to get all the equipment. Has relieved Capt. Thomas of squadron command. Go to town with Penn [*Lt. Pennington H. Way*] in afternoon. Band concert by artillery band.

FIRST TO BOMB

At this time, the 96th Aero Squadron had yet to have any observers or pilots assigned to it. Rath remained as an instructor with the 7th A.I.C. Major Harry M. Brown [1890-1960], West Point 1914, took command of the 96th Aero Squadron on April 17th from Captain George C. Thomas [1873-1932] who became the Adjutant of the 96th. Brown learned to fly in 1916. He arrived in France with the 1st Aero Squadron in September 1917. In January 1918 he assumed command of the 12th Observation Squadron. However, he preferred bombers and managed a transfer to the command of the 96th Squadron.[17] At 27 he was old for a pilot, but squadron commanders did not often fly in combat; so his age was not really a concern for anyone other than himself.

The 96th Aero Squadron had been formed in August 1917 at Kelly Field, Texas without aircraft or flying personnel. The squadron, under the command of Captain Thomas and consisting entirely of support personnel, went through basic training at Kelly Field. On October 9th the squadron was ordered to the Aviation Concentration and Supply Depot at Mineola, N.Y. for duty overseas. On October 27th they sailed from New York aboard R.M.S.S. Adriatic arriving at Liverpool on November 10th. They immediately were sent to the 7th Aviation Instruction Center at Clermont-Ferrand, arriving November 16th.

At Clermont-Ferrand, they took charge of the hangars, armament and ground transportation. The mechanics learned the intricacies of the Breguet bomber and Renault engine at the nearby Michelin factory where the Breguet 14 B2, the aircraft to be used by the 96th Aero Squadron, was built. The men of the 96th were, in the spring of 1918, thoroughly trained as practical aircraft mechanics.[18] Their knowledge extended far beyond maintenance to include repair of battle damage as well as repairing crashed aircraft.

April 18

My birthday [*age 33*] but nothing to do. Take a walk and write some letters.

---

[17] Ruffin, Steven A., "Major Harry Brown and his 'Lost Flight' of the 96th Aero Squadron," *Over the Front*, vol. 19, no. 3, Fall 2004, p. 196-221. This article extensively covers Brown's final flight.

[18] Gorrell, Series E, vol. 14, p. 65.

FIRST TO BOMB

April 19
 Learn that headquarters wants to make artillery observers out of us. What next? Every body is in an uproar. Still waiting orders.

April 20
 Rain. Band concert and opening of new Y [*YMCA*] hut. Artillery band plays good.

April 21
 No orders yet. Take walk with Major Rader [*Major Ira. A. Rader, West Point graduate. From November 1917 to September 1918, he commanded 7th A.I.C. He was attached to First Day Bombardment Group in September 1918*]. They are keeping up the night sessions in our room. Last until 3 in morning.

April 22
 Write letters all day. Yesterday all the officers haul ashes and fix up the new hangars. French people don't know what to make of it. [*illegible*] yesterday evening. Got a letter from the Y [*YMCA*] today. It was a funny one.

April 22 – 28
 More days of loafing. When will we ever get out of this place. All the squadron men are up in the air. We hear that we are to be turned into reglage [?] (artillery observers) after all our bombing training. Next we hear that we are to be chasse [*pursuit*] observers. Major B. [*Brown*] goes to headquarters to see what he can do. Wish I could have gone to England to take up night bombardment. Rains all week. Take a few walks. Play hand ball and quoits for the first time. Heard some very surprising news about the [illegible]. Major returns but no decision so far as to our future. Have a talk with the commandant & Ellis. Perhaps it will all work out all right. Lt. Strobbing [?] and I go to C-F [*Clermont-Ferrand*] Sunday.

April 28
 Hear band concert. Have good dinner. German big drive towards Ypres and the coast still continues no decision to the battle so far.

Don't think I will get home this summer. Capt. R. [*Reed?*] stages a big party early in the morning. Capt. C [*unknown last name*] proud and I am a soldier of France.

April 29 – 30

Still no news. Loaf around camp in morning. Go to town in afternoon. Run across Lt. Handley [*?*] and Dr ____ in the montreur [*?*]. Think they are trying to make a date. After dinner talk to Madame Soulin [*?*] and others in the store [*?illegible*]. All of us go to the cinema. Afterwards walk home takes me just an hour.

April 30

Try to play ball. guard stops us at line. Go to town in the afternoon. Run in with Jimmy McBride. Run across "Chic" Evans [*Lt. Elisha E. Evans*]. Go to station. Have dinner with Chic and other Lt. see major B, [*illegible*] and Cody. Chic and I almost stay in town with Lt of Artillery who buys [*brings?*] horses. Get last car by running.

May 1

No news. New major in camp to stay. Major B comes in at noon. Play quoits with Van Sickle Steve. Steve can't overlook the money. He gets my goat by his crabbing. [*illegible*] Lt. leaves for front. Pay day. Get letters from home also write up in University Club news [*University Club Bulletin, vol. 2, No. 1, April 1, 1918; contains short article: "'Judy' Rath Into It"*]. Walked out from C.F. [*Clermont-Ferrand*] with Capt. Thomas [*Capt. George C. Thomas, Jr.*] in afternoon. Concert at "Y" [*YMCA*] in evening. Party in room afterwards Major Narden [*?*] arrived in camp.

May 2

No news yet. Have to help censor mail. Captain Keel [*Reed?*] has a boil on his nose. Go to C.F. in afternoon almost walk to Royal. Walk home alone from M.F. [*?*] Sit around and talk in room.

The graduates of the 7th A.I.C. bombardment training program and the support personnel of the 96th Aero Squadron waited, seemingly endlessly, while rumors swirled around them, wondering

FIRST TO BOMB

what their duty assignment would be. During this period boredom and frustration reigned; morale was low.[19] Weeks passed until the men, pawing the ground and chomping at the bit, were referred to by their comrades as the "bewilderment squadron."

May 3

YMCA secretary takes French lesson at table [*illegible*] and [*illegible*] he is a scream. Lots of mail yesterday. Heard from my old friend Marion Lehutty [?] She isn't engaged any more.

May 4

Penn Way taken suddenly sick at night Dr. Marshall [*Lt Frank Marshall, surgeon*] sore when he is brought to room. Take walk with Capt. Thomas. Hear all about roses also hear we may get new planes when we go out. Steve get peeved at Chic Evans at quirts. Go to town afternoon Gundy [*Lt. Andre H. Gundelach*] & Chick. Go to Royal with Chick Evans for dinner. Run into Smithy [*probably Lt. Herbert D. Smith*] at dinner. The gang in room when I get back.

May 5

Heninger [?] wakes me at 7 o'clock to police camp. Lucky I didn't stay in town. Rains all afternoon. Read and sleep all day.

May 14

Same place – nothing to report. We are all getting stale. Go to town in afternoon after censoring mail. May 6 ran across Tom Norton on the street. He has been here a month. C.A. [?] instructing in artillery "mechanical maneuvers" – whatever that is. Lt. Metzger [*Lt. Arthur R. Metzger*] and Robertson [*Philip Robertson*] killed May 10. Everybody in a gloom – first accident in the camp. Metz seems to stall the machine 150 ft. from the ground. After plane hits ground it burns up. Has been raining ever since we came back from [*illegible*]. Stay up to Royal and walk around for first time. Country folk dance [*illegible*] of [*illegible*]. Eat meal in town mess out here is getting to be tired. Still in room with two majors.

---

[19] Thomas, *The First Team*, p. 24.

FIRST TO BOMB

Chapter 6

96th Aero Squadron

May 14

Today we hear that orders are coming through for us to go out. Go to town in evening. McBride, Red [*Pennington H. Way, also known as Pinky*], Chick Evans, eat dinner with Capt. Reed, Chick, Red & Steve. Run across Tom Norton again walk around and go to picture show – picture taken in L.A. [*Los Angeles*]. Ride home Capt Reed mocking birds singing.

Rath makes no diary entry for May 15th. However, it was on this day that he was assigned to duty with the 96th Aero Squadron. At last the delay was over. Excitement swept through the waiting men. In the days that followed the rest of the original flying members of the 96th Aero Squadron received their orders.

May 16 – Thursday

Tom [*perhaps Tom Norton or Lt. Thomas Farnsworth*], Steve, I walk up to farm house with people from the Magasin de peche [*fishing shop*]. Drink milk. Horribly long walk.

May 17 - Friday

96th goes out tomorrow. We go up for a flight in afternoon. Codman [*Lt. Charles R. Codman*] breaks a machine – it will keep us back for several days. Most of the fellows of the squadron go in town to sleep. I walk out at midnight.

FIRST TO BOMB

May 18 – Saturday

96th goes out at 3 in the morning. Buy some supplies during the day. Get caught in the rain on trip back. Big dance in evening. 65 nurses and artillery band. Met Miss Fairlamb – what a name. Van Sickle and his midnight walk. This place is getting hot.

Well before dawn Major Brown and most of the ground crewmen of the 96th left, by vehicle, for Amanty airdrome, 225 miles to the northeast, near Gondrecourt. Rath did not go with them. On May 22nd Rath, the aircraft, and flyers will take off for their new home, the airfield at Amanty.

May 19

Sunday – walked with Major Rader. Went to town in evening. Took a walk. Couldn't get a bed in Clermont – went to Royal – got one by saying I was an American officer.

May 20

Had a good restful sleep last night. Came out to camp. Went to Royal and picnic in afternoon. Lt Van Sickle. French jealous because French girls go with Americans. Entente reception in evening. US gen'l makes a fine speech. Lt. Thibaud [?] takes us to a show afterwards. I sleep in town.

May 21

Sellers [*Lt. Cecil G. "Swede" Sellers*] gets wire from Major Brown. Looks as if we go tomorrow. Big dust storm: Major Lufberry [*Raoul Lufbery, 94th Squadron Ace, shot down May 19th*] has been killed.

The 96th Aero Squadron flew on May 22, 1918 to their operational airfield at Amanty, located 12 miles north of Neufchateau, 4 miles northeast of Gondrecourt. The squadron's aircraft were 20 worn Breguet 14s that had been used by the students at Clermont-Ferrand. The aircraft had been reconditioned but had seen a lot of hard service.

Figure 5. Leaving Clermont for Amanty, May 22, 1918.

    The Breguet 14 B2 bomber was fitted with a 300 h.p. type 12 F.E.V. Renault engine powering a 9.5 foot propeller. This French plane was designed to be both a trainer and a bomber. It had flight controls in both the forward and rear compartments.

    The Breguet 14 B2 entered service in the summer of 1917 and was considered a very sturdy, dependable machine. About 5,500 of them were built during WWI. The frame was duraluminum, which was lightweight and very strong. In addition, the wheels and tail skid had shock absorbers allowing for harder landings without damage. It weighed 3,900 pounds and could carry 500 pounds of bombs. The top speed was 110 mph, and it had a service ceiling of 19,000 feet and could climb to 6,500 feet in 7 minutes and reach 17,000 feet in thirty minutes. The usual operating altitude for bombing missions was 12,000 to 15,000 feet. [20] The combination of speed, long range, durable construction, high altitude climbing power, bomb weight load carrying capacity and maneuverability made it an excellent bomber. Its twin Lewis machine guns provided adequate defensive capability. As the squadron was to learn, this defensive capability was only effective when the aircraft was flown in large, and tight, formations and best when the formation was accompanied by pursuit aircraft.

    An unusual feature of the Breguet was the automatic flaps across the trailing edge of the lower wings. These flaps would depress downward when the velocity of the aircraft was low providing more

---

[20] Gorrell, Series E, vol. 14, p. 74.

FIRST TO BOMB

lift, thus allowing for a 10-15 mph slower landing speed and a slower take-off speed. At normal flying speed, the flaps would automatically raise. Thirteen Sandow cords, much like bungee cords, on each lower wing supplied the energy to move the flaps.[21]

Figure 6. Insignia, guns and observer's window.

A variant of the Breguet 14 B2 was the A2 known as the Corps d'Armee. It was designed as an observation aircraft and did not come from the factory equipped with bomb racks or bombsights. These had to be modified in the field. Nor did it have the observer's windows on either side of the rear cockpit. It also did not have the automatic wing flaps. During the four months of operations, the 96th squadron received thirty-one B2s and thirty-two A2s.[22]

---

[21] Toelle, Alan D., *Windsock Datafile Special, Breguet 14* (Berkhamsted, Herefordshire, Great Britain: Albatross Productions, LTS), 2003, p. 48.

[22] Gorrell, Series E, vol. 14, p. 77. Toelle, *Breguet 14*, p. 19.

FIRST TO BOMB

May 22

Left C-F: 7:20 in leading machine with Hooper [*Lt. Thornton D. Hooper*]. Map works fine – our squadron gets off in fine style. Run into a little rain over mts. Hoop misses Dijon but I discovered it. Gundy [*Lt. Andre H. Gundelach*] & Tom Farnsworth [*Lt. Thomas H. Farnsworth*] pass us at Dijon. Our squadron goes down at Dijon for oil & gas. French fliers give us lunch. Great bunch. Get to Dijon 9:15 leave at 11. Pat Anderson [*Lt. Charles Patrick Anderson. Lt. Hugh S. Thompson is his observer*] turns back soon after leaving Dijon – engine trouble. Some map – we hit the route right along. Pass Neufchateau and I find field for Hoop [*Lt. Thornton D. Hooper*]. Land at 12:10. Gundy & Farnsworth beat us in.

Our quarters are better than expected. All the rest come in afternoon except Pat Anderson [*and his observer, Hugh S. Thompson*]. Think he went to Dijon. We fix up our barracks. Try to locate band concert – walk for miles through woods. Go to bed early get good night's sleep.

May 23 – Thursday

Pat Anderson engine on bum. Sends for mechanician. Rains today – movies in YMCA.

May 24 – Friday

Gundy makes another sensational landing. Hooper, Smith, Codman, Steph [*probably Lt. Robert G. Stephens*] & I go to Neufchateau in afternoon. Run into Roger Clapp. Have bath & dinner at Lafayette Flying Club. Come back in crowded camion [*truck*].

May 25 – Saturday

Hooper and I take a ride in plane. Bumpiest ride I have had. Stop in at the 94th for lunch. See Rickenbacker [*Eddie Rickenbacker, 1890-1973, American Ace*] & Elsie Janis [*American singer, songwriter, actress, 1889-1956*]. Great brunch of fellows. Move over to other bunk house in evening.

FIRST TO BOMB

May 26 – Sunday

Get late break fast. Eat lunch in officers mess – first meal. Take walk in afternoon to Amanty with Cawston [*Lt. Arthur H. Cawston*] Col. [*blank*] comes in evening. Bad news about planes. Probably more latrine rumors.

May 27

Work around barracks in morning. In afternoon walk to Gondrecourt. Throw my knee out on way back. Have a bad night.

May 28

Sad news – hear Lt. Stearns [*Lt. William S. Stearns, died May 25, 1918*] & mechanician get killed down south. Play ball in afternoon. Lt. Clapp comes over in afternoon. Party in evening Red [*Pennington H. Way*] says good chance to win money if – Go to movies but leave as it isn't any good. Sleep on boards – give Clapp my bed. Offensive starts up again [*American offensive, battle of Cantigny*].

May 29

Take Clapp over to C in auto. Wonderful trip beautiful day. Write letters in afternoon. Fellows go to trenches. Gundy discourses on why he won't marry. Try to go to Paris tonight.

May 30

Drive towards Chateau-Thierry on. Took 24 hrs to go to Paris on train. Passed any number of troop trains met two ferry pilots. Very near freeze trying to get some sleep. Get into Paris 8 o'clock at night. Take a walk. Shell from big gun explodes near by - hits M-. Alert sounds for air raid. I go down metro subway at Place de la Opera - quite a sight. Heavy barrage all clear sounds at 1:30.

May 31 – Friday

Shop in morning and afternoon. Meet John Pike and Reagan [*Bill Reagan*] at six. Sure was good to see them again. John and I go to dinner and a movie in evening. John has an adventure – his French is just as funny as ever. Alert sounds as we are in taxi but we get home in time.

FIRST TO BOMB

June 1 – Saturday

Shop in morning. Run across Mounds in afternoon. Same old boy. John Pike, Bill Reagan and two other Lts. hire a cab and ride over to the Latin quarter where we have dinner. The major general banquet – the mix up. The alert sounds stumble down stair way - get cab but he won't take me. Terrible barrage no traffic moving. Find taxi by roadside. He says he will take me after raid, so I sleep in the taxi – he says I'm a hard one. I finally get home at 2:30.

June 2 – Sunday

No noon train. Go up to Pike's room. Get bath. Wonderful day. In afternoon we go for a ride in a cab go to Bois - every body seems to be there. I have an adventure. Almost miss my train. Sad farewells of soldiers go to front. Am in compartment with lots of French officers. Get some sleep.

June 3 – Monday

See anti air craft bursting around Bosche. Get to Gondrecourt at 8 in morning. Remember funny Lt. of infantry in train. Nice French Lt. of infantry who is going to change over to aviation. Play baseball in afternoon – rotten. Clapp is in squadron now. Movies in evening.

June 4 – Tuesday

Fool around in morning. Sleep in afternoon. Walk to nearby town in eve.

June 5 – Wed

Red Way is sick. Write letters. Hear in evening that we are to go over line on Monday with English Major leading. I am to go as our majors observer. Everybody enthusiastic. Go to bed early.

June 6 – Thurs

Go in a formation flight today. Hooper and I lead. Go to Chaumont hangar, Vittel & Neufchateau. Flight goes very good. Only one machine drops out. Pretty cold at 14,000 feet.

FIRST TO BOMB

June 7 – Fri

English major [*Major Alexander Gray*] & Capt [*Captain Ward*] arrive. They are to lead us over on first trip. Another group goes on flight today. A [*illegible*] bursts into Farnsworth's machine and busts it up.

June 8 – Saturday

Go up as major's formation in morning. Formation pretty ragged. He is a good flyer and the trip goes fine. Run into some clouds and had to fly pretty low. In afternoon Hooper & I lead other formation. Clouds very low & air very bumpy. Two pilots get lost. Difficult flying – visibility poor. I don't like the new formation as well as the old. Hope they stick to the old one. Sellers and Chick Evans go over to fly with the English. Pretty bored tonight – poor movies.

The squadron was continually trying out new methods of operation, especially new formations. There was much discussion, and experimentation, in determining how many aircraft should be in the formation and where they would be positioned relative to each other. This was critical for obtaining a good bombing pattern but also for defense against attacking aircraft. Yet, at the same time the formation had to be maneuverable. Communication between aircraft was difficult being confined to hand signals, wing movements and, when enemy aircraft were seen, the firing of flares from the Very pistols.

Major Harry M. Brown, commanding officer of the 96th, was eager to get into the fight. All Spring he had been chomping at the bit for this moment. However, after being at Amanty for three weeks, which had been spent on topographical orientation flights and formation flying practice, he was becoming impatient to begin bombing in earnest. Unfortunately, he faced a major problem...the squadron had no bombs.

There were three types of bombs used in WWI. High explosive bombs had a large ratio of weight of explosive to weight of the bomb container. Most of the bombing missions of the 96th carried these bombs as they worked well with railroad terminals, railroad tracks and marshalling yards. They were excellent for buildings of all kinds. Fragmentation bombs carried small charges of explosive in a heavy steel shell that would fragment to inflict casualties on unprotected personnel. Incendiary bombs were used for setting fire to ammunition dumps, buildings, airdromes and even grain fields.

Pilot David Young recalled that anti-personnel, incendiary, and high explosive demolition bombs were sometimes loaded in the same bombload as they did not want to miss the opportunity of setting fires or causing casualties. He described the crater from a demolition bomb to be "about the size of the average room."[23]

The tangled supply bureaucracy could not get bombs from the depots to the 96th Squadron. Officers from the 96th Armaments Department roamed the countryside in the squadron Cadillac visiting French supply depots seeking a transfer of bombs to the 96th. At one French air park they found a supply of bombs. They returned to Amanty for a French-speaking officer to make the formal request to the French commander. Lt. Hugh S. Thompson retuned with them to the supply depot but was unable to persuade the French to let him have any bombs. The squadron commander, Major Brown, followed up with a personal visit and multiple pleas. The result was that on June 8th, 1000 115-mm short Michelin bombs were delivered from French Air Park No. 13 located back at Clermont.[24]

These small anti-personnel bombs weighted only 8.1 kilos each, or 18 pounds, and they were unsuitable for attacking railroad yards, but they were at least bombs. Naturally, high explosive bombs would have been preferred but the Major was commanding a bomber squadron and rather than wait he would bomb with what he had. The 96th Aero Squadron was now ready to take the war to the enemy.

June 9 – Sunday
Not much to do tonight. Sit around camp.

June 10 – Monday
Still waiting for something to do. Do a little work with maps.

---

[23] Leiser, "Red Devil in a Breguet, David H. Young, 96th Aero Squadron," p. 160.

[24] Thomas, *The First Team*, p. 35 and Gorrell, Series E, vol. 14, p. 79.

FIRST TO BOMB

Chapter 7

The First Bombing Raids

June 11 – Tuesday
    Hear we are going over tomorrow. Get my maps fixed up. Have conference in evening. Gundy is sore because he can't go over. Go to bed about 12 but have hard time getting to sleep.

June 12 – Wednesday
    Do a formation in the morning. The major and I lead. Bad visibility and we almost get lost in the clouds. Sight doesn't work very well. Get ready to go over at 3 o'clock in afternoon. We are to bomb Dommery Baroncourt [*Dommary-Baroncourt*]. Big crowd watches us get off. Pictures etc. Get off in good formation – 8 machines. Two turn back first half hour. Clouds very bad – we are soon above them and can hardly see the ground. Once in awhile get glimpse of ground. Takes us about 2 hours to get to Verdun. One more machine turns back here. Anti aircraft begins to bump us as soon as we get over. I think Bosch machines start after us as soon as we get across. They appear higher than we are but far in the rear. Just as we reach our objective they catch up with us and as soon as we drop our bombs we begin fighting. They keep swooping under our tails and we keep shooting to drive them off. We fight all the way back to the lines. Pat Anderson gets two bullets in his plane and certainly the Bosch gets their fill as we can see our tracer bullets go straight at them. Anti air craft pretty accurate. Glad to get back of the lines. Verdun certainly

is all shot to pieces. We take a long tour getting back home. Two machines run out of gas but make good forced landings.

The whole camp had given us up for lost. Mechanics begin to look over planes for bullets. This is the first bombardment trip made by the Americans. The major (H.M. Brown) and I led it and everything went OK as some of the observers saw bursts on the RR tracks. Almost too tired to sleep.

Figure 7. Participants in the first bombing raid plus Royal Flying Corps spectators. Left to right: Newberry, Tucker, Lewis, Beverly, Duke, Capt. Ward (Royal Flying Corps), Maj. Gray (Royal Flying Corps), Strong, Rath, Maj. Brown, Mellen, Browning, MacDonald, Tichener, Smith, Ratterman, H. Thompson, and Evans.

The June 12, 1918 bombing mission is important. It was the first ever bombing raid conducted by an American squadron. The honor of dropping the first bomb was Lt. Howard G. Rath's as the lead observer. He was also responsible for navigation to, and from, the target. Procedure called for all observers to watch the underside of Rath's airplane and when they saw his bombs drop, they would drop theirs as well.

Each aircraft carried sixteen of the small bombs making a total weight of 129.6 kilos or 285.7 pounds for each aircraft. 1428.5 pounds, a bit more than ¾ of a ton was the total weight of bombs dropped in the first raid.

FIRST TO BOMB

The pilots and observers that dropped bombs in the first mission were:

    Major Henry M. Brown
    2nd Lt. Howard G. Rath
    1st Lt. Durwood L. MacDonald
    2nd Lt. Alfred R. Strong
    1st Lt. Joseph M. Mellen
    2nd Lt. Rowan H. Tucker
    1st Lt. Henry C. Lewis
    1st Lt. Claxton H. Tichener
    1st Lt. Charles P. Anderson
    1st Lt. Hugh S. Thompson[25] [26]

The route to the target from the home field at Amanty was never as the crow flies (see Figure 8 on next page). The formation would always stay on the allied side of the lines for as long as they could to avoid flying over enemy territory as much as possible. On this mission they flew forty-five miles NNW to Verdun. From Verdun they turned NE, crossing the lines, and flew fifteen miles to the target at Dommary-Baroncourt, then returned by way of Verdun, for roughly thirty total miles over enemy-held territory.

None of the airmen, other than Rath, had been across the lines or been fired on by anti-aircraft or attacks from enemy aircraft. The bombers encountered both anti-aircraft fire and was attacked by German pursuit planes which were fought off by the observers. Poor weather and perhaps poor navigation caused the flight to be much longer than anticipated. Some aircraft ran out of fuel before they could reach Amanty. However, all returned safely to a celebration of the first mission.

Despite the raid being only a pinprick and a minor annoyance, the 96th Squadron, and hence the American Air Service, had gone to war. The news of this mission was widely reported in American newspapers. The New York Times article appeared on the June 15th

---

[25] Gorrell, Series E, vol. 14, p. 85.
[26] For biography of Lt. Hugh S. Thompson see Harrington, Hugh T., *Destiny's Wings, Four Months in Day Bombardment: The Story of Lt. Hugh S. Thompson, 96th Aero Squadron, U.S. Army Air Service in World War I*, Gainesville, Georgia: privately published, 2019.

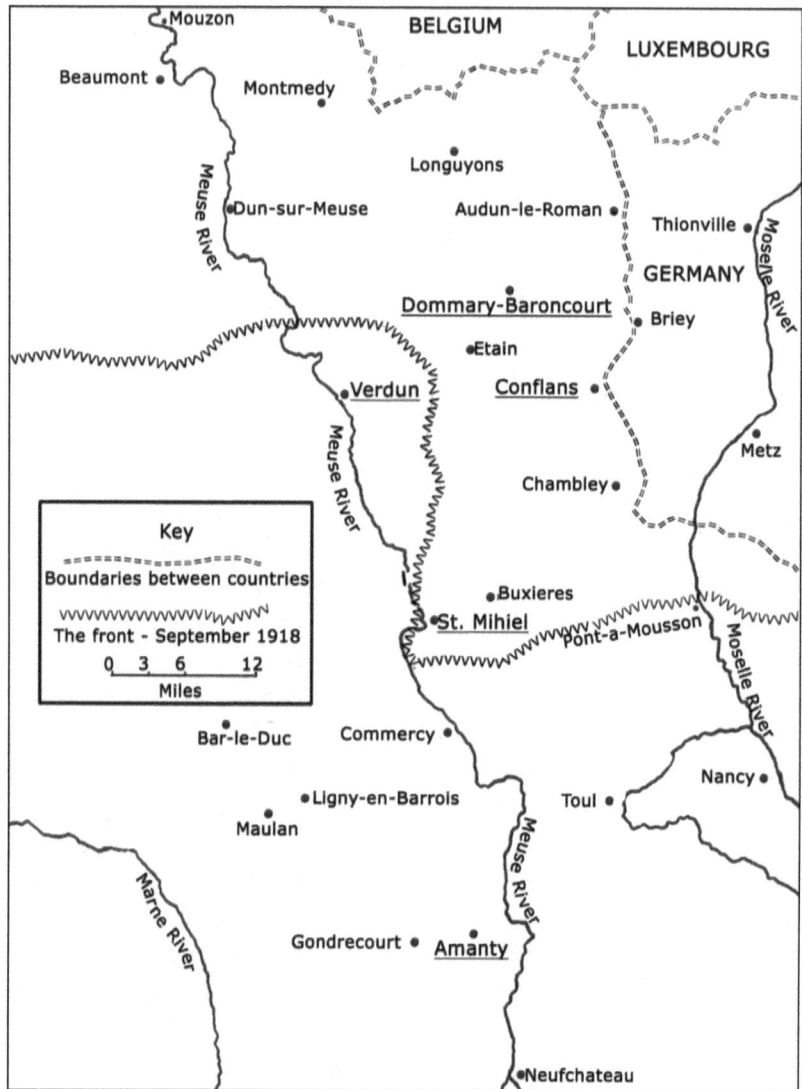

Figure 8. Map of operational area for the 96th Aero Squadron

front page under the headline: "Americans In Air Raid Beyond Metz Bomb German Towns and Railway: Fight Way Back." Neither the squadron, nor the men involved, were named. However, the Americans had taken the offensive in the air, and the newspapers were *making the most of it, even if their reporting was not strictly accurate.*[27]

Howard Rath wrote a typewritten letter to his brother Walter on June 13th the day after the first American bombing raid. Although it is not part of the diary it is included here as it provides another look at this historic mission. It has never been published.

June 13, 1918
Dear Brother,

Yesterday we put on our first show (bombardment) and those of us that got to go along are pretty happy. Everybody wanted to go along and help inaugurate this new American express train service into Germany but of course we all could not go.

The Major of our squadron flew the leading machine on this trip and as he took me along as his observer, it kept me pretty busy before the trip getting my maps fixed and making all arrangements.

You can imagine how the whole camp was sort of on "its toes" before the trip and when we left the ground it seemed as if one side of the field was black with people. We got off in good shape and began to make a certain tour in order to get our altitude. After flying about fifteen minutes two machines signalled [sic] that they were going back, and we learned when we got back that one observer had made quite a hero of himself for part of the engine hood had blown back cutting a hole through the wing and was just going to cut the tail off, when the observer leaned way out of his cockpit and grabbed and held the whole hood piece away from the tail until they landed.[28]

But to go on with our trip. We hadn't gone very far before we began to run into clouds and they got worse and worse and soon we were traveling right over a bank of clouds and there wasn't a sight of the ground anywhere. It made a beautiful picture to see the machines

---

[27] "Americans in Air Raid Beyond Metz Bomb German Towns and Railway; Fight Way Back," *New York Times*, June 15, 1918, p. 1. "Fliers Damage Depot; Escape Enemy Planes," *The Sun* [New York], June 15, 1918, p. 1. "80 Bombs Dropped in First Air Raid, Americans Defeat German Planes After Expedition," *The Sun* [New York], June 15, 1918, p. 1.

[28] As it is unclear which airmen took off on the mission, only those who completed the mission are positively known, it is impossible to determine who the heroic observer was.

sailing along in formation over the clouds. It was just like traveling over a field of snow, but we were not out to make a pretty picture and it didn't help much in getting to our objective to have nothing to steer by but the compass. As luck would have it after about half an hour the clouds began to take on a Swiss Cheese effect and I steered by what I could see through the holes.

By the time we got to the lines we were going good and the clouds were getting farther apart. But the Bosche put the wind up on us (anti air craft-shrapnel) as soon as we got over, and their shooting was mighty good. Then, too, just as we passed the lines, I spotted three Bosche planes (two single seaters and one double seater) bearing down on our rear at a higher altitude and far behind and gave the signal to the rest. All the time that we took making our objective the archies were booming away and the Bosche planes were gradually creeping up.

Being in the leading machine I had plenty to do with directing the course toward and over the objective, sighting for the bomb dropping, giving the signal to the others to drop their bombs and trying to keep an eye on the Bosche planes at the same time.

As soon as we had dropped our bombs on the objective we turned back towards our lines and as by this time the Bosche were within shooting distance, all of us got busy with our machine guns. The Bosche began swooping down, trying to get under our tails and we kept scrapping all the way back to the lines. Just as we saw our tracer bullets streaking right at them, they all turned back.

Our trip had been a pretty long one and when we got that line behind I felt easier because I knew we were getting pretty short of gas. We began to hit the clouds again as soon as we got across and because of them we made rather a long tour getting home. Two machines ran out of gas and had to make landings about 30 miles from our camp but they both made good landings. The rest of us made the home field and it was some relief to climb out of the old bus.

The mechanics discovered that one of the rear machines had a number of bullet holes in it but the rest of us never got a scratch. From the bullets we got we reckon the Bosche planes have plenty of souvenirs to remember us by. From what the rear men saw of the bomb bursts we consider the raid was a success and today we were

FIRST TO BOMB

willing to sit back and let the other group do their trick. If things go right we will make a tour again tomorrow.

>   Love to all,
>   Your brother
>   Howard

The diary continues with the entry for June 13, 1918.

June 13 – Thursday
    Sure glad to loaf today. Other flight is to go over this afternoon but had to be given up on acct [*account*] of machine trouble.

June 14 – Friday
    Get up late - good sleep. Hooper and I do some machine gun work and also try the sights for the bombs. Looked as if flight was off but they leave at 5:30 to bomb Conflans. Gundy and Red Way lead. 6 machines start five get across. Anti air craft but no Bosch. They claim they got the round house and hit the RR tracks. Get back in about 2 hours.

June 14 Friday [*Chronology is confused. June 14 was Friday, but the events of that day were described in previous entry, also dated June 14, Friday.*]
    Bad day. No chance to fly.

June 15 – Saturday
    Still bad. Dud weather

June 16 – Sunday
    Go on formation flight – very bumpy weather. Trip comes off good. Don't feel very well. Liver I guess.

June 17 – Monday
    Go to Neufchateau. Get bath – also batteries for sighter [*bombsight*]. Commissary was closed. Get couple letters today.

FIRST TO BOMB

June 18 – Tuesday

Sure feel rotten today. Weather still looks bad. Take walk around and see how sighter [*bombsight*] is getting along. Play cards in afternoon. About 3 weather clears up – go out to field. Major says thinks we will go on a show. Hooper and I are to lead. Get away at 6 o'clock. Go by way of Joinville [?] and Barb Dehre [*Bar Le Duc?*]. Pass into clouds at Verdun. Make turn so Farnsworth can catch up but he falls back again. Formation is OK especially Gaylord [*Lt. Bradley J. Gaylord*] & Summersett [*Capt. James A. Summersett*].

Not much archies as we go over – just enough to let us know they see us. Visibility very poor – can't see much ahead.

Hooper can't see objective. I spot it just as we are coming out of the clouds. Steer Hooper over RR yards of Conflans and drop bombs just as we reach center. See Gaylord's drop also. Try to see bombs drop but too hazy – field glasses don't help. Archies keeping popping all way back to St. Mihiel – some break very close. Get good view of trenches. Weather is better now. See camp from lines. Get home at 8:20. Learn two observers jumped the gun and dropped their bombs in the woods. Gaylord's & one other machine have shrapnel holes quite close to Pressler [*Lt. Warren S. Pressler*]. Major says we may go again tomorrow.

June 19 – Wednesday

Had a rotten night liver seems to be bad. Just get into break fast in time. Raid is on today if weather is good. At lunch hear that Hooper and I are to go over to English camp this afternoon for 3 days. Sellers & Evans are to come back. Told Major I would rather stay and puke [?] and he thinks it is a good joke.

Pack up after lunch go over in Cadillac. Smith goes along. Have tea at five o'clock. Rain starts in after dinner – rains all night.

June 20 – Thursday

Take a look around the hangars. Chinese coolies – English sergeants. Sleep in afternoon. Rotten weather. Raining again. Make out "who should we notify slips" again.

FIRST TO BOMB

June 21 – Friday

Another "dud" day – rain. I sure feel rotten. Hooper and I take a look around the hangars and take a walk. Major Gray & Capt. Ward come back in evening. Say that Major Brown has the men standing by for a raid.

June 22 – Saturday

Rains in morning. Go down to hangars. Take a walk in afternoon. Some English colonel and American Capt. are here for lunch.

June 23 – Sunday

Still dud. Nothing doing. In afternoon it clears slightly and all squadrons leave for raid. One machine crashes in trying to land with dead stick. Formation came back about 8:30 pm. Very cloudy and could hardly see objective. Only one Bosch seen.

June 24 – Monday

"Dud" day again. Doesn't look as if we will get to go on raid. The DH-4s go on raid. Have a thick time over Metz. 99 [*99 Squadron, Britain's Independent Force*] loses some men. Nap [*Thornton D. Hooper's nickname is "Nap" or "Napoleon"*] & I go to Nancy in afternoon. Get good bath, good meal at Millers [*Walters?*]. We look over the town in the evening.

Nap is very funny. Major Grey says we shouldn't hurry & we don't. Stay in Nancy all night.

June 25 – Tuesday

Nap telephones to Capt. Thomas about coming for us. We go back to camp in English mess truck. French autos collide. Dud afternoon but English. Go to Metz again. Bad clouds over objective. 99 loses some machines and two observers. 55 loses one.

June 26 – Wednesday

Major Gray says truck will come for us today. We get already to go home but car doesn't show up. All squadrons go for raid on Karlsruhe [?]. 36 start 29 get across lines. Wish I could have gone along. Bad weather again. 55 [*55 Squadron, Britain's Independent Force*] loses one machine – 99 several. Bad archies and some E.A.s

FIRST TO BOMB

June 27 – Thursday

Last night the German night bombers were around Nancy – plenty of archies & some flares. Tried to get Achey [*archey?*] but missed. Cadillac comes for us at 2. Have lunch at Nancy. Get home at 5. Raid called off because of "dud" weather. McChesney [*Lt. Harold A. McChesney*] now flies with major. Seems good to be back home. Lots of mail. Movies.

June 28 – Friday

Just loaf around camp. Have a meeting and tell fellows what we learned over with English. Go to bed early.

June 29 – Saturday

Go to Chaumont with Capt Thomas & Col. R. [*perhaps Col. Ira A. Rader*]. See negro troops. Major starts on raid but they all come back – too cloudy. Get good news about new planes. More mail. Summertime game.

June 29 – Saturday

Lectures this evening and lots of mail.

June 30 – Sunday

Fool around camp. Pick berries, write letters and get some mail.

July 1 – Monday

No show so far. New motors are on the way. Hoop goes to hospital to get rid of the itch.

July 2 – Tuesday

Motors come and they hope to have planes ready for a show on the 4th. They look good.

July 2 – Tuesday [*two diary entries for July 2, Tuesday*]

Motors are being put in. One is ready for test in evening but too cloudy. I don't feel just right as yet. Names posted for show. Major is going to take McChesney.

FIRST TO BOMB

July 3 – Wednesday
    Gundy takes one motor up. Others are being finished fast. Turns cloudy in afternoon. Rained in evening. No show tomorrow. Cold and chilly but at last have a stove. Major goes to Toul [*headquarters*].

July 4 – Thursday
    Bad rainy day. Sit around barracks in morning. Play in officers ball team in afternoon. Make two star catches. Movies at night. Very good. Major Grey [*Major Alexander Gray, M.C.*] & Capt. Ward fly over. Major Brown almost goes to Toul.

July 5 – Friday
    Putting motors in machines. Looks as if we will have plenty of planes soon. Roger Clapp and I walk over to Amanty. Walk over to town in evening.

July 6 – Saturday
    Fool around hangars in morning. Feel tired. Go to sleep in afternoon. Some one wakes me to tell me Roger Clapp & mechanic Dunn [*Sgt. Robert J. Dunn*] crash and burn up in plane. First loss in Squadron & Roger was one of the first fliers I got to know in the squadron. Movies at night. Red Way & I stand watch from 5 to 7 in morning.

July 7 – Sunday
    Rogers funeral at "Y" I act as pall bearer. Jim [*probably Capt. James A. Summersett, Jr.*] reads my paper. Japanese priest says last mass.
    I go over to see Knapp Hooper [*Thornton "Nap" Hooper*] in afternoon.

July 8 – Monday
    Knapp [*Nap*] comes back to camp. Penn Way, Gundy, Joe Mellen [*Lt. Joseph M. Mellen*], Pat Anderson, Bob Thompson [*Lt. Robert E. Thompson*] & Rat [*probably Lt. George A. Ratterman*] go to Nancy in mess truck. Penn & I go out to skating rink. See major [*Maj. Harry M. Brown*] in evening. He tells us about his army life –

also that he is going to promote us. Bombs drop close by. Lots of anti aircraft.

July 9 – Tuesday

Rains in morning. M.P. tells us where to find barber curfew. Have lunch with Major Brown & Lt. [*illegible*]. Go over to English squadron to see Liberty motor. Major goes to sleep in car while we wait in field. Have tea with Major Grey. Get back for late supper & movies (he loves the weird princess).

FIRST TO BOMB

Chapter 8
Major Harry M. Brown Flies into Infamy

July 10th was a memorable day that brought the squadron a great deal of unwanted notoriety. Chief of the U.S. Air Service, William "Billy" Mitchell, called the failed mission of that day "...the most glaring exhibition of worthlessness we had had on the front." And, "I know of no other performance in any air force in the war that was as reprehensible as this."[29] [30]

The weather on the 10th was bad, too bad for flying a mission. Flights were cancelled due to high wind, rain and heavy clouds obscuring the ground. Then, late in the afternoon, Major Brown ordered a raid, with six aircraft, on the railroad yard at Conflans fifty miles to the north. The wind was blowing strongly from the southwest.

July 10 – Wednesday

Sit around the hangars in morning to talk to Jim [*Capt. James Summersett*], [*illegible*], [*illegible*], Pat [*Pat Anderson*], Ratt [*Ratterman*] & Knapp [*Lt. Thornton "Nap" Hooper*]. Play ball in afternoon. Major decides to start out on raid in spite of clouds. Leaves at 6:10 pm. with 6 machines. At midnight none of them have returned. At 7 o'clock the sky was completely covered with clouds. At eight it

---

[29] Mitchell, William, *Memoirs of World War I*, (New York: Random House, 1960), p. 242.

[30] The best account is Ruffin, Steven A., "Major Harry Brown and His 'Lost Flight' of the 96th Aero Squadron," *Over the Front*, vol. 19, no. 3, Fall 2004, p. 196-221.

FIRST TO BOMB

was raining. There was a storm coming up in the S.W. when he started and a quarter of the sky in the north was covered with a black cloud mantle. The best thing that we can hope is that they became lost and landed in France. All during the night we listened for motors although it was foolish.

July 11 – Thursday

Clear but windy that is only white clouds in sky. We all hung around hangars all morning expecting the machines to come in. Didn't want to waste time shaving as was afraid that I would miss seeing them come in. In the flight were Major Brown, Lt. McChesney, MacDonald [*Lt. Durwood L. MacDonald*], Strong [*Lt. Alfred R. Strong*], Mellen [*Lt. Joseph M. Mellen*], Tucker [*Lt. Rowan H. Tucker*], Smith [*Lt. Herbert D. Smith*], Ratterman [*Lt. George A. Ratterman*], Browning [*Lt. Robert C. Browning*], Duke [*Lt. James E. Duke, Jr.*], Lewis [*Lt. Henry C. Lewis*] and at noon no machines have returned and no news. Headquarters has asked all squadrons on the front to look for them. Right after lunch Swede Sellers and I hear a plane and rush out to the field thinking it was a Breguet but it is only a Salmson going by. The wait is getting on all our nerves. At about 4 o'clock an intercepted wireless of the Germans reports that "out of six American planes that tried to attack Coblenz five have been captured with their crews." How did they ever get way up there when they were going to bomb Longuyon [?] or Conflans. Capt. Thomas & Lt. Leroy [*Lt. C.H. Leroy, Intelligence Officer*] go to Toul to report to General Foulois [*Benjamin D. Foulois*]. Take weather report with them – General Foulois only asks for a description of the weather. Says squadron will go right on new planes to come – new teams also and that Major Dunsworth [*James L. Dunsworth*] will be sent to take command. Captain Summersett takes charge of post in interim. We hear about six o'clock that other plane has come down in Germany but don't know how or anything about crew. Everybody is blue – 4 men out of our barracks. Movies bring [*illegible*] no pep [*illegible*].

None of the 6 planes came back. They were all captured. After the war the captured men gave statements to the Air Service about their last flight. Pilot Lt. Durwood MacDonald stated that the flight left Amanty at 6:05 pm and flew on "a very changeable course between east and north from 6:30 p.m. to 8:00 p.m. without seeing the ground at any time." At 8:00 they came to a "big opening in the clouds" and "saw a large city below. Made a complete tour of the city, without dropping bombs and started for home at 8:30 p.m. I started my own course, southwest, and at 9:45 p.m., I was forced to land, owing to darkness and not having any more gasoline."[31]

In his statement Major Brown admitted that he was blown off course by winds that were far stronger than the 12 to 22 miles per hour he had been led to expect from the pre-flight weather report. His observer, Lt. Harold MacChesney, measured the wind speed with his bombsight to be 65 mph.[32] The entire flight was forced to land as they ran out of fuel attempting to return. The flight overflew Conflans and was blown 120 miles north-east of Conflans to Coblenz where they attempted to return, hopelessly flying into the wind.

It must be remembered that the American airmen were inexperienced in the art of bombing and had a great deal to learn. Flying under adverse conditions, especially poor or even non-existent visibility, was a lesson that would come back to haunt them with a vengeance again during the St. Mihiel offensive in September.

July 12 – Friday

Lost my leather coat yesterday on flying field. Somebody must have it but doesn't turn it in.

No more news until news when German verify report that sixth plane was shot down. The poor fellows evidently were trying to get back when they went west. Spent all day packing up Joe Mellen's effects. How we miss the fellows. Nap [*Hooper*] reads part of Mac's [*either Lt. Harold MacChesney or Lt. Durwood L. MacDonald as both were lost on this mission*] love letters - he certainly knew a lot of girls. Joe's letters are most all from his fiancée, she writes every day. Nap [*Hooper*] and Charlie Codman go to Paris this evening to fix up Mac's and Smithy's affairs. Penn Way and I walk over to village nearby [*3 illegible words*] because in reading some letters that Smithy's mother wrote he discovered that they think he is exaggerating his experiences to his wife. Certainly it is better not to enlarge on any experiences one

---

[31] Gorrell, Series M, vol. 10, p. 183-4.
[32] Gorrell, Series M, vol. 10, p. 34-35.

has. The French chef turns out a good mess. A big mail comes in but I don't draw today.

July 13 – Saturday
　　Potter around barracks in morning. Ride to Gondrecourt with Penn Way & Steve [*probably Lt. Robert G. Stephens, Supply Officer*]. Buy papers, American communique of the 11th says, "as a result of a bombing raid last evening 5 of our planes are missing." Short and bitter. Get a couple letters at noon. Feel better. Burn letters in afternoon and then walk over to Amanty to get a bath but it is full of enlisted men so walk back across the fields. Wonderful afternoon. Get a letter from Elizabeth at dinner time. There is an amateur performance at the "Y" tonight. Penn, Steve and Chick [*Lt. Elisha E. Evans*] sing some trios. As one enlisted man told Chick afterwards "It wasn't Chick's fault as the other fellows didn't stick with him." The steel guitar players were good and made me homesick. Went over to Capt. Thomas' hut and had a long talk with him about the fatal bombing raid.
　　Swede [*Lt. Cecil G. Sellers*] and I expect to go to Nancy tomorrow. The French Bastille 4th of July.

July 14 – Sunday
　　Lt. Sellers and I go to [*blank*] for the celebration. Ride over on truck with enlisted men. Run all around town trying to get a shave. About as many American flags as French out. American band concert in afternoon. All population is there. Band stand covered with American flags. Run across lots of the English crowd in the afternoon. They said that our accident had helped them in that they didn't have to go on bad day any more. Swede and I run across some body who knew Major Brown. Every body seems to know of him.

July 15 – Monday
　　No more news so far. Hot and sultry all day. Sleep some of the time. Rains in evening. Capt. Thomas calls me over to see what I think of statement he has prepared re the catastrophe. It is all true and yet I am the only one whose name is mentioned and all the testimony seems to be mine. Capt. Summersett comes in and takes a hand in criticism. He evidently doesn't like the evidence. Before we get

through our new major (Dunsworthy) [*Maj. James L. Dunsworth*] arrives and I go home.

"Nap" [*Hooper*], Penn, Sellers, Chick Evans and I go over to village after concert.

Concert is a funny affair. One singer was good but he got to ragging [?] the crowd and it didn't take "I can make more noise than the whole crowd." Our own party makes up for it. See a wonderful display of rockets and flares at the front on the way home. Also hear German planes bombing and see anti aircraft.

July 16 – Tuesday

Hot again today. Sleep a good deal as there is nothing else to do. Get a few letters. Get word that the new offensive is on at Chateau Thierry again. The Americans seem to be doing well. Good movies in evening.

July 16 to Friday 19

Not much doing for days. We are still marking time. One night a German plane comes over and bombs a near by town. Can hear the plane circle around and hear the bombs explode. Beautiful moonlight nights. We borrow a gramophone and keep it busy playing out in the moonlight. The Americans are beginning to get in some good licks around Chateau Thierry. Wonder how the Germans like that.

July 19 – Friday

No news. Start ball a rolling to get leave. Have papers sent to Toul but miss major.

July 20 – Saturday

Hooper, Gundelach, LeRoy, Capt. Thomas and Major Dunsworth. Gundy says Penn has been over lines more than I have and hit the target – I wonder where he gets that misinformation. The result is that Gundy says he won't play unless he can have his own way. The outcome is that after I had been mentioned by Capt. Thomas first for Chief Observer that they decide we are both good men and so I am made Asst Operations Officer (whatever that is) and Penn is made chief observer. C'est la Geurre. I go over to Toul in afternoon with Charlie Codman, Penn Way, Val Hower [*Lt. Virgil H. Hower*]. Pretty ride. Have dinner with two American girls one is Miss McMahan who draws cover designs Saturday Evening Post. Very clever girls. Went to movies in evening. Didn't sleep very well. Caught a mouse.

July 21 – Sunday

Raining and I have the blues. Take a long walk with Hooper in afternoon. Go to E [*possibly village of Epiez-sur-Meuse, 2.25 miles NE of Amanty airdrome*] and in a Cafe have lots of fun talking to the "landlady's daughter." Hooper and I discuss various things. G [*presumably Gundelach*] in particular. Have a long talk with Major Dunsworth [*1887-1956. Rath is 2 years older than Dunsworth*]. He asks me my opinion on several matters. [*"Having"? or illegible*] a meeting with LeRoy and Penn. Run across LeRoy over at E [*possibly Epiez-sur-Meuse*] – very strange.

Church and movies in the evening. Very good. Jim tells story about the monkey. Mrs. C. [*unknown last name*] is there. I was going to Paris this evening but decided not to go. Thompson [*Bob or Hugh*] and Pressler [*Lt. Warren S. Pressler*] go. Pressler certainly did me a great favor.

July 22 – Monday

Hooper and I get up early and bomb sight was broken but we make some good hits – best made so far. Come back at 10 and go to observers school. Penn does very well. Expect to go to Paris [*illegible*]. There are 7 planes at C [*Clermont-Ferrand?*] for us. Pilots are going over to fly them back.

Chapter 9

The Squadron After Major Brown

July 22 – Monday [*Rath repeats the date & day*]
Hooper and I bomb in morning. Make best record so far. Sight didn't work very good. Have talk with Penn Way in afternoon. Wonder if he takes right attitude. Catch train to Paris in evening.

July 22 Monday to July 26 Friday
See John Pike and Bill Reagan. Both of them are making good. John has been recommended for 1st Lieut. Seems good to be in Paris again. [*illegible*] in aviation are coming along good now. Run across Breen.

July 26 – Friday
Take morning train out of Paris. Walk out to camp. Nothing doing while I was gone. Go to movies in evening.

July 27 – Saturday
Offensive of Germans is going the wrong way. Trained American troops are pushing them back at Chateau Thierry. Go to movies in evening and have feed [*?*] afterwards. Lots of mail.

July 28 – Sunday
Rain today. Study some. Write letters. Go to movie in the evening.

FIRST TO BOMB

July 29 – Monday

Study down at intelligence office. Looks like we will soon make a raid. Go to movie in evening.

July 30 – Tuesday

Go to observers class in morning. Listen to lecture by Gundelach. Play handball in evening. Good game. Americans are doing good work at Chateau Thierry. Bosche comes over in night. Can hear him drop bombs and shoot machine guns at Epiez. Believe he was looking for us but misses.

July 31 – Wednesday

Hooper and I take a flight on a cross country trip. It doesn't go very well. Someone shoots at us near Bar-le-Duc. In afternoon Major Dunsworth explains what he wants me to do as Operation officer. I get out first operation order. Gives me something to do any way. We expect to go on a show tomorrow.

August 1 – Thursday

It is a good day. Gundy takes his formation on a cross country in the morning. At 3:30 in afternoon Hooper and I lead off the first formation for a "show" at Conflans. Gundelach and Penn Way lead the second formation. Poor visibility. Lose track of Gundy but pick it up at Verdun. Formations get mixed up – some one shoots out into Germany alone but comes back in a hurry as there is a German plane out there. Have a hard time to see Conflans. My goggles break and the light in the sighter burns out. It is hard to see through the sighter. Can only see Conflans just as we get over it. Have to pull Hooper around in a steep bank to get him over it. [*Pilots are "steered" when near the target by the observers, who can see the ground below the aircraft through a slot in the floor and pull reins attached to the pilots' shoulders.*]

Pull the bombs and as luck would have it make just about perfect hit. Possibly because everyone was practically stalled on a bank. See about a dozen bursts right on the railroad track. Many others among the machine shops and warehouses. We must have hit some ammunition stored because there are several big explosions and a number of fires start. Best work we have done so far.

FIRST TO BOMB

On way home anti air craft is thick and very accurate. We go over two airdromes in order to get pictures. Get back to camp at 6:20 p.m. The formation flying was very poor. Possibly because it was the first time we tried two formations. Everybody is tickled over our bombing and say it is the best one so far. Gundy is sore about the flying altho he congratulates me on the bombing. Make out my reports and see part of a movie. Major Dunsworth asks "me" what to bomb tomorrow. We are going to Dommary-Baroncourt if it is a good day. Major talks over with me who will fly.

August 2 – Friday

Bad day - rain no show. Find that our pictures taken on the raid are N.G. [*no good*]. Too bad as they would have been a good example of what can be done in precision bombing. [*illegible*] reports out. Work in office. Go to Neufchateau in afternoon with major. Gundy breaks bone in his hand while fooling in a truck. We get caught in the rain on our way back to camp. Jim sets us up to chocolate.

Forgot to say that Charley Codman smashes into the road while trying to make the field yesterday with his bombs still on. Lt. Hower his observer has his arm broken.[33] Machine is wrecked.

August 3 – Saturday

Hoped to put a show on today until last minute but weather isn't good. May go tomorrow. Work in office almost all day. D.H. 4 with Liberty motor comes in camp to play with 8th squadron – first appearance here of the Liberty. Go to movie in the evening. Afterwards have a feed and talk fest in our "front" room. Beaucoup letters today.

August 4 – Sunday

Bad weather. No flying. Write letters. Go to meeting & movies in evening.

---

[33] Hower suffered a compound fracture of the arm which put him out of action for the duration.

FIRST TO BOMB

August 5 – Monday

Bad weather. No flying. Go to Neufchateau with major & Jim [*"major & Jim" crossed out*] Chick Evans. Try to locate Miss Lillian Wilson at hospital. Can't be done. See Lt. Hower he is out of the game for some time. Run across Johnny Cooper from Los Angeles. He is in Red Cross work. Got lots of news from him. Have a feed in the evening.

August 6 – Tuesday

Bad weather. Fool around in morning. Get a few letters. Hoop, Chick, Gundy, Penn and myself go to Epiez in afternoon. Try to get Suzanne to cook a chicken dinner for us but she can't buy any chickens. Learn that French and Capronis [*Italian bomber*] are coming to E [*Epiez*]. Trip back through woods seems very long. Play game and Steve and O'Toole [*Lt. James A. O'Toole*] get sore. Big feed afterwards.

August 7, Wednesday

Doesn't look very good for a show. Lt. Blakeman surprises us by walking into our barracks. Says people say about our squadron "We have met the enemy and we are theirs." He is going to England to start a Handley-Page School. Was dark all day - couldn't get over. Went to Toul in afternoon. Just bit rainy. Look like drive [*sentence not completed*].

August 8, Thursday

We expected it to be clear this morning but clouded up again. Hoped to get across some time during day but no luck. Am now chief observer while Way is gone. Observers play pilots ballgame in afternoon and we win. Play volley ball in evening and we lose. Movies in evening. Penn and Gundy are in Paris on leave. Looks as if it might be clear tomorrow.

August 9 – Friday

Still cloudy. No raid. Play volley ball in evening. Great game. Another [*Amateur?*] show afterwards. Capronis go out on bombing raids afterwards.

FIRST TO BOMB

August 10 – Saturday

Bad news for me. Nap Hooper, my pilot gets ordered to report to Tours as C.O. of a squadron. He certainly deserves it. Everybody is glad of his raise. Just what is to become of me is a problem! It looks as if observers have as much chance as a baby girl in China – sort of necessary evils. Go on a formation flight with Capt. Summersett. Try a large formation of 11 - it doesn't work very well. Go down to depot with Nap in evening. Sure hate to see him go. He is going to see if he can take me along.

Capronis carry more freight during the night into Germany. We are going to try two flights in the morning. Lt Sellers & Evans lead one – Lt. Gaylord & I the other – six in each.

August 11 – Sunday

At last a fair day. We get away on raid about 10:30. We have a hard time keeping up with first formation. At lines as only 3 are left of our formation we join on with first formation. Sellers & Evans lead us too far north. After wandering around for about 20 minutes we bomb on signal a little R.R. station probably north of Longuyon. Very poor bombing – only two bombs hit track as far as I could see. Instead of turning west and coming right home – we wandered south past Dommary (our objective) Conflans even to Vigneulles. Finally turned west and got out after get shot up. We beat it right home as soon as we got across lines (Gaylord & I). I tried the double control. It was pretty hard to do much from the backseat. [*The Breguet aircraft had dual controls allowing it to be flown from the observer's rear seat.*]

Farnsworth and Sellers had to land because they ran out of gas. Everybody was pretty sore about the trip. Chick Evans was there with the excuses.

Raid for afternoon called off. We try to locate town we bombed but can't locate it. Work until 11 making out new operation picking out formation etc. Are going to try out Lt. D.H. Young [*David H. Young, pilot*] & Lunt [*Samuel M. Lunt, observer*] as formation leaders tomorrow. Going to try Conflans at eight o'clock. Two raids if possible. I won't go on this raid myself. Before going to bed I notify all the men.

FIRST TO BOMB

August 12 – Monday

Hoped to get two raids in today but only got in one. Lt. D.H. Young and Lunt lead. Went to Conflans and had good results. They dropped the bombs across eastern neck of RR tracks. Four planes were hit by archies at Conflans.

August 13 – Tuesday

Weather keeps good. Gaylord and I lead a formation of 7 over Dommary-Baroncourt. We get in a big wind drift and our bombs hit eastern [*"eastern" crossed out*] southern neck of tracks. Capt. Summersett sees a shrapnel burst under our plane & thinks it is our bombs so drops his and 3 others follows him. [*Bombing procedure called for the formation to drop bombs when the lead observer dropped his bombs.*] About 5 seconds later I drop my bombs on the tracks and the other 3 follow mine but it is too bad that Capt. Summersett & the others dropped their bombs too soon. Didn't run into much anti aircraft as we came back by way of Etain. Get word that Hooper has asked for me as operation officer. I talk it over with major [*Dunsworth*] and decide to stay here with the 96th as operation officer.

Sellers gets word in evening that he is made C.O. of a new bombing squadron. He and I are going to lead a raid tomorrow to Longuyon.

About this time an incident occurred that to some degree illustrates that the Americans were still learning their trade. Rath, returning from a mission, discovered that he could not see the white windsock at the field. Normally, fliers would land into the wind, and the wind sock indicated the direction of the wind. Without the wind sock they used the smoke from a farmhouse chimney to give them wind direction. Upon landing they discovered that a unit from the Engineering Corps, connected with security, had visited the field and painted the wind sock with camouflage paint so "German fliers could not locate our field!" That was immediately corrected.[34]

---

[34] Rath, Howard G., "Origin and Development of American Aerial Day Bombing During World War I," unpublished typescript, p. 13.

FIRST TO BOMB

August 14 – Wednesday

Get an early breakfast and Sellers and I start leading formation off of the ground at 8:30 for Longuyon. 9 planes leave ground but only 5 go across lines. It is the first time I have ridden with Sellers and also the first time any of us has made a raid on Longuyon. So it has a keen edge on. It takes us 1 ¼ hours to get to Verdun on account of stiff north wind. Good day and have no trouble finding objective but Sellers thinks it is farther on. He is afraid that we will run out of gasoline so won't bomb the way I want him to. The wind makes us drift and although several bombs hit tracks a number of them hit to west of depot. On the way back 3 German chasse planes [*illegible*] us up and we fight at long range to lines. Instead of going straight back to erdun [*Verdun?*] Sellers runs way down to Vignuelles.

In afternoon Tom Farnsworth and R.E. Thompson [*Lt. Robert E. Thompson*] lead a formation to Dommary but they drop the bombs short of the target. No E.A. [*Enemy Aircraft*]

Pinky Way & Gundelach come in camp in evening their leave has still 4 days to run but they have run out of money. They are rather surprised to see Hooper gone and Sellers leaving. Pinky wants to go on a raid in the morning. Major says Gundy can't go before he sees the doctor and at least not tomorrow morning. Pinky is going up with Alexander. Capt Thomas tells me that major is trying to do all he can to see that I get a squadron also has recommended me for a promotion! (All favors gratefully received) Movies had a bosche plane this evening.

FIRST TO BOMB

Chapter 10

The War in the Air Heats Up

Until the second mission of August 15th, the 96th had been enjoying the excitement of the war, but their own adventures had been somewhat less than death-defying. Theirs was a strange, detached war on the periphery. They, of course did not fly every day. However, on flying days they would pass over the lines into enemy territory, observe the scorched earth of the trenches below, and be fired at by usually inaccurate anti-aircraft fire. The few enemy pursuit planes they had seen had kept their distance and were not a mortal threat. The 96th had not really been in the heat of serious action. Theirs was the life of an exciting, though not especially life-threatening, flying club. That was about to change as the tempo amped up to the coming mid-September nightmare.

August 15 – Thursday

Alex [*Lt. Arthur Alexander*] and Pinky Way get their formation off at 10:25. Six planes get across to objective Dommary. Make a good hit on central and southern part of track. Eight bursts on tracks. Not much archie and only one plane at a distance.

In afternoon Gundelach and I lead a formation over to Conflans. Start at 4:30 p.m. with eight planes. Only five get across. Gaylord [*Lt. Bradley J. Gaylord*] – Pressler [*Lt. Warren S. Pressler*], Codman [*Lt. Charles R. Codman*] – O'Toole [*Lt. James A. O'Toole*], Young, C.P. [*Lt. Cecil P. Young*] – Anspach [*Lt. Ralph Anspach*], Alexander [*Lt. Arthur Alexander*] – McLennan [*Lt. John Charles Earle (J.C.E.)*

FIRST TO BOMB

McLennan], [*illegible*] ourselves. Go north from Verdun and come down R.R. tracks over Dommary to Conflans. I take my sight time last time, over Dommary. Sight doesn't show up well. We make a perfect hit in the middle of the tracks at the western end of the yards just where the railroad tracks narrow down into the 4 tracks going to Metz. I afterwards drop 2 special bombs on the round house. Only one burst of shrapnel over Conflans. Conspicuous by its absence. It was so quiet that I was shooting up the hangars at Piereriux [?] when I thought I saw something black drop out of our machine. Looked over the other side of the machine and discovered it was one shrapnel burst but it was [*evidently? - illegible*] to spot us for Bosche chasse machines for a bunch of them were coming around from the front on our right. There were eleven of them altogether as some of them came over the top of the formation. I leaned over to ask Gundy if they were our protection and when he shook his head I got busy with my guns. Just about the same time every one else opened up on them to and what a roar of machineguns there was. Six Bosche dropped behind but five got under our tails in formation and fought with us in that position all the way back to the lines. I saw Gaylord's machine slipping around and I thought he was going down but he was only maneuvering. Gundy turned into the sun for us and went west to the lines, we, whaling away all the time. The brace wire on our rudder was shot away and just then the machine turned slightly north and for a moment I thought we were out of control. The bosche left us at the lines but just before that at [*Wolf?*] one of them came up along side slipped off on a wing as if he had been hit.

Codman crashed landing on the field and Gundy discovered he was shot in the foot. Our rudder pulleys had been shot away and we had nine holes in our plane altogether. All the planes were badly shot up. One # [*blank*] C.P. Young had 6 shots through the propeller. Everybody was pretty excited. Could hardly settle down to sleep.

Twenty-three-year-old Lt. Andre H. Gundelach was the squadron's most experienced pilot and a powerful personality. At 19 he had enlisted in the U.S. Navy as a boatswain's mate, making a round-the-world cruise. He requested a discharge from the Navy in 1916 so he could join French aviation. He attended French flight schools from March to July 1917. He was assigned to Escadrille SPA

FIRST TO BOMB

95, a pursuit squadron, at the Front. While with SPA 95 he shot down his first German and was awarded the French Croix de Guerre with Palm. In September he requested a transfer to Escadrille Sop. 111, which was a bombing squadron. He flew Breguet 14 B2 aircraft on bombing missions until December 1917. He was transferred to the 7th Aviation Instruction Center at Clermont in January 1918 as a bombing instructor.

His operational experience flying the Breguet made him the senior combat flyer when he was assigned to the 96th Aero Squadron on May 17th. Pilot Charles Codman wrote that Gundelach handled a Breguet as if it were a pursuit plane, which inspired confidence among the other airmen.[35]

Figure 9. Breguet 14 B2.

---

[35] Gordon, Dennis, *The Lafayette Flying Corps: The American Volunteers in the French Air Service in World War One*, (Atglen, PA: 2000), p. 190-191.

FIRST TO BOMB

August 16 – Friday

Young, D.H. and Lunt lead a formation of eight off the field at 4:55 pm for Dommary. Gundy & Penn go along in position #2. They miss the target a mile – have heavy archie. Gundy's 18 [*aircraft number*] gets a great big tear in the fuselage – no Bosche. Get a letter from home.

August 17 – Saturday

We couldn't have any flight today too cloudy. Play volley ball in morning. Go to movies in evening. Chick goes to Nancy with others and almost gets shot down by an Ally plane. He also hears that the Richtoffens circus has been sent here to get us. [*Manfred von Richthofen (born 1892, KIA April 21, 1918), a.k.a. Baron von Richthofen and the "Red Baron." Despite his death, his pursuit wing of four squadrons, informally known as the flying circus, had a reputation for combat excellence.*] Going to try to have a show at 5 in the morning for the first time. Get a great letter from H.C. [*unidentified*]. Movies in evening.

August 18, Sunday

We were to go on a raid at 5 o'clock this morning. We all got up at 4 o'clock - got a light breakfast and went out on field. None of the machines were out. Someone forgot to wake the mechanics. Capt Thomas rushes around and gets things going. The weather is bad but the flight leaves the ground at 5:30. I go back to sleep. Fellows come back at 6:30 as it is too cloudy to go across.

Fool around in Red Cross hut in morning. Write letters in afternoon & sleep. Go to services in evening. Write to H.C. [*unidentified*] then go to bed.

August 19, Monday

Bad day – no raid. Am sent over to C [?] by the major as they asked to have a bombardier sent over. Didn't know what I was to do until I got there – it was to drop American bombs out of a Liberty D.H. 4 with Wimperis sight. Generals, colonels, majors etc over there to watch the show. Lt. Prince [?] from the 94th piloted the ship.

No one seems to know what is to be done or how to do it. Several try to explain the sight to me but none of them agree. It is my first ride

FIRST TO BOMB

in a Liberty. Takes no time to get our altitude but we can't get over the target. They put the target right next to the road – if a bomb should hit the road they would have to build another one. Then too they sit in their machines on one side of the target – Get [*guess?*] they have confidence in my bomb dropping. I drop the first bomb at 3000 ft – it hit on the opposite side from the grandstand and about 500 feet off – the direction is good. The reins that I fixed up work fairly well.

I have a hard time getting the next bomb off as it won't release. It finally gets off and drops a little better than the first but practically in the same place.

We go back to the hangars. Mechanics have left and no one seems to know about dropping the other bomb. A colonel finally drives up and wants to know why we don't bomb. We explain – get action and finally get off with a lopsided bomb caused by the bomb on one side. Drop this bomb correcting for wind but it is worse than the others. They all practically drop in the same place. There is something wrong – probably the engine speed. Get some supper and come home. Miss out on the presentation to Capt. Thomas [*Capt. Thomas is being transferred*]. He goes into a vrille [*tailspin*] from [*when*] he sees his presents. Everybody in the barracks is in N_.

August 20 – Tuesday

Bad weather – stay around camp. Clears in afternoon. Capt. Thomas leaves us today – everybody sorry. Raid at 4:45 pm. Gundelach and Way – clouds – they miss target. Beautiful night all the night planes are out. I am homesick. Go on raid tomorrow with Gundy.

August 21 – Wednesday

Wonderful day. Leave ground at 10 o'clock. Wonderful visibility. Can see for miles. Start out with nine planes 3 drop out. Not much archie. Can see Longuyon from lines. We crab across objective but get a good hit at east end of yards. Try some trick dropping on army afterwards but fell a little short. No bosche seen. Just a fine joy ride. It is really warm in the air. Some mail today – that helps some. Got a letter from Han [?], also from the folks. They have received my letter of June 13 regarding our first raid.

FIRST TO BOMB

In afternoon Gundelach & Penn Way lead raid to Audun-le-Roman. They miss objective. No trouble. Get some more letters tonight. Wonderful moon. Another raid tomorrow but I'm not on it.

August 22 – Thursday

Another hot day. D.H. Young and Lunt [*Lt. Samuel M. Lunt*] lead formation to Conflans. Archie very hot. They didn't do extra well. Some hits on tracks. 450 cars on siding. Brad Gaylord and I were going to lead formation over in afternoon but it got too windy. May go tomorrow. Movies tonight but no good.

August 23 – Friday

Still windy. Gaylord & I lead formation over to Conflans at 10:30. Clouds and wind. Brad [*Lt. Bradley Gaylord*] has hard time getting altitude. Gundy & Penn have to turn back. I get very nervous as visibility is very poor. Verdun doesn't look right. Bevo [*perhaps pilot Lt. Belmont Beverley*] flies in front of formation at Dommary Baroncourt. We have a bad drift on. Have a hard time holding on target. C.P. Young pulls bombs on my signal to get ready. Lots of them drop in field. Make about 7 hits on tracks in front of station. One hits warehouse. Have to buck wind back.

In afternoon Gundy & Penn take formation over. Miss Conflans all bombs drop in fields. Gundy is sore. Their plane gets hit on bomb rack. Clouds up at night.

August 24 – Saturday

Rainy. No shows. Go to Neufchateau with Chick and Charley Codman. Don't go to Nancy in the afternoon. I don't feel very good. Last evening Way, Hooper & Swede Sellers came in. Their camp is near by. Poor movies tonight. Penn, Pat [*Pat Anderson*] & T. Farnsworth have a party today. Three observers come in today [*the list of commissioned personnel that appears in Gorrell, Series N, vol. 16, p. 181-184 does not list any observers being assigned near this date*]. They are infantry observers. Have never bombed.

August 25 – Sunday

Gundy & I lead formation to Conflans. Get started at nine thirty. Rather cold. Somebody shoots red flare just as we reach lines. I get

FIRST TO BOMB

my wind up [*get anxious, nervous*] as I think we're in for a long flight. Have a hard time getting my time on my sighter. Forget to shoot pistol [*Very pistol*] to get ready to bomb. Drop bombs and others follow. Make a great hit – all but about four drop on tracks. One drops right beside the round house. See no E.A. but archies sure are accurate. Pressler gets his suit ripped up the back. Make trip in practically 2 hrs.

In afternoon Penn Way & D.H. Young lead formation to Longuyon. Their machine got hit with shrapnel and they were forced to drop out of formation while they were in Germany [*German occupied France*]. Got back alright. D.H. Young was sitting right on top of shrapnel shell case. It had gone right thru the seat. Farnsworth & R.E. Thompson take lead. Miss Longuyon a mile. All get home OK. Rotten landings 18 [*aircraft #18*] gets wing bent. Didn't get to church today. Gundelach & I are going to lead formation to Conflans tomorrow.

Howard Rath, and virtually all others from the top commanders to the airmen themselves, believed that they were doing significant and useful damage on their bombing missions. This was not true. The Germans had teams of workmen at the railroad yards ready to repair damage, or install switches to reroute traffic around blast damage, as soon as the attacking aircraft had left. Often damaged track could be repaired within an hour of the attack. The unknown reality was that before the attackers had landed at their home airfield and reported their bomb hits, the damage had often already been repaired.[36]

August 26 – Monday
    Dud day. Volley ball & letters – silhouettes in evening.

August 27 – Tuesday
    Dud day. Volley ball & movies. Hear that Germans have started a day bombing squadron.

---

[36] Suddaby, Steve, "Aerial Bombing in World War I," *Over the Front*, League of WW I Aviation Historians, vol. 35, No. 4, Winter 2020, p. 320.

FIRST TO BOMB

August 28 – Wednesday

Very windy – no raids. Go to Gondrecourt for money orders. Have dinner at Bidouville with Pat [*Lt. Pat Anderson*], Tom Farnsworth, Chick and Penn. Spring chickens & wonderful potatoes. Hear that we are to do hazardous work soon. Looks like good weather tomorrow.

August 29 – Thursday

Hoped to get off a raid all day but weather still held bad. Are going to try an early morning raid tomorrow. Went to Nancy.

August 30 – Friday

Gundy called me at 4 o'clock – Had just got to bed at two o'clock.

We got into the air at 5:15 a.m. Still dark. Everything shows up best I have ever seen. Hills, rivers, R.R. Lots of movement on R.R. Pretty cold up in air. I get taken short and have terrible time. We have to buck heavy wind from north. Intended to bomb Conflans from north but wind swings us around to south. Nine planes get over. Heavy barrage at Conflans. I have a hard time with sighter and can't see the lines on the lens. We undershoot the whole town. Three Bosche leave Peuxux [?] hangars but don't come very close. Shoot at balloon on way home. Get home in time for breakfast. We plan other raid for eleven o'clock. Penn & D.H. Young lead. Were going to Longuyon but wind too strong hit Dommary instead. Make good hit. In morning raid Charley Codman has his engine hit with shrapnel [*illegible*] planes back to Columbey.

Gundy and I lead 3rd formation for day to Conflans at 5 o'clock in evening. Have hard time getting our altitude. Gundy plays a trick on me. Only six planes out of ten go over. Pat [*Anderson*] and Tommy [*Farnsworth*] go over practically alone. Bomb from north – or rather crab from west to east and drop our pills right along station and across the tracks. Archies aren't so bad this time. See big [*illegible*] burning on American side on way home. Biggest day we have had so far.

English plane with sergeant pilot & Lt Alsford observer come down on our field with bad engine. They were on their way to bomb Luxemburg. I put up Lt. Alsford in our barracks for the night. He seems to fear flying. Doesn't look very good for a raid tomorrow. Get a good look at Metz today. Think makes my 12th raid.

August 31 – Saturday
    Bad weather no raid. Take bath in afternoon. English get their DH 9 fixed and start off at 4:30. Sgt pilot makes several banks at low altitudes. Gets to spinning and crashes at east of field. Tank explodes & afterwards one of bombs explodes. It is terrible. Pilot is blown into bits. Observer Lt. Alsford's killed & broken up pretty badly too. A D.H. Liberty [*DeHaviland D.H. 4 with American Liberty engine*] crashes same way earlier in the day & kills pilot.
    Pat, Fanny [*perhaps Thomas H. Farnsworth*], Penn, Gundy, H. Thompson [*Lt. Hugh S. Thompson*], R.E. Thompson, Chick & I go over to French town for dinner. Duck & wonderful potatoes. Movies afterwards Fatty Arbuckle & Mable Normand also Bill Hart & Rory.

September 1 – Sunday
    Bad day. Major takes me over to [*illegible*] to meeting of majors. Has to do with reports. Get dinner at [*illegible*]. Stop to see Swede [*Sellers*] & Nap [*Hooper*]. Nap got into a spin yesterday. C.P. Young serves us a wonderful dinner. Rains all night.

September 2 – Monday
    Good day. Penn & Gundy lead formation to Audun [*Audun-le-Roman*] but miss target. Come back over Conflans. No Bosche but archies are rather bad. In afternoon D.H. Young & Lunt were to lead formation but he doesn't show up so Brad Gaylord & I lead formation to Longuyon. Get in a bad wind & drift & miss target. On way back four bosche attack us they circle around from front and get under our tails. I shoot 450 rounds almost running out of ammunition. We had to buck the wind on the way back and it seemed as if we didn't move at all. Once I thought one Bosche was going down in flames as he began to show black smoke and went into a tight spiral – but I believe he came out alright. Poor concert at the Y in evening.

September 3 – Tuesday
    Early morning raid - reminded me of getting up to go duck hunting. Got out on the field at 5:15. Gundy and Penn were to lead

raid but came back with bad motor. Tried it the second time but motor went bad again. D.H. [*Young*] & Sam Lunt take lead. Go to Longuyon run into bad clouds – some snow – make a good hit on machine shops & R.R. yard. Pat Anderson and Lt [*Hugh*] Thompson bombed Dommary because they couldn't make Conflans.

In afternoon Gundy & Penn Way lead formation to bomb Conflans. Slip by station but bomb lines running out. Make some good hits. Penn thinks they hit bridge. They had a bad fight with 3 Bosche coming back. Think one was brought down. None of our planes were hit. Poor shooting by the bosche. They come around the side again – poor flying. About [*illegible*] [*illegible*] Poor pictures at the Y. Go on a raid tomorrow morning if the weather is good.

September 4 – Wednesday

Pretty good day – but about 50 mile gale at 5000 meters. Gundelach & I lead a raid off at 10:50 for Dommary Baroncourt. Had a hard time getting no. 1 [*aircraft number 1*] started. Gaylord & Chick as deputies start off but we catch up with them and take their place or rather our own place. Nine planes start but Gaylord & Chick have to come back on account of not making altitude. Gundy goes pretty far west and he can't believe Verdun is so far east. We finally get straightened out however and go across at about 4000 meters. Not much archie. I say to Gundy – this sure means Bosche planes on account of wind we go to Conflans instead of Dommary Baroncourt. Everything seems to be lining up fine and I think that we are going to make a great hit but we make a terrible miss. Only way I can account for it is that on coming on to the objective I steered Gundy to the south and that we ran into some wind that held us back. Our bombs dropped short.

As soon as we left Conflans Gundy spotted some Bosche and so I fired the signal. They closed in on us at Trierville [?] (6 Pfalz scouts) and how they did shoot. One Bosche kept right in back in the middle of our formation and kept shooting at our plane. Our tail was in the way and it was hard to get a shot at him. Finally had to do something so in shooting shot our rudder braces on one side away. Just at the beginning of the fight one bullet came through my seat (I was standing up) with an awful crack but didn't hit me.

FIRST TO BOMB

All the observers were banging away at a great rate. By this time four other Bosche came around the side to join in the fight. As one came circling around from the front I let him have both guns and I am sure every other observer on that side did the same (R.E. Thompson [*Lt. Robert E. Thompson*], Lt. Kelley [*probably Lt. Arthur Kelly*] & Lt H. Thompson [*Lt. Hugh S. Thompson*]). The Bosche stood it for awhile and then began to go down. Hugh Thompson says he saw him crash. The others kept right on us and things looked very serious. I could see Alexander's [*Lt. Arthur Alexander*] & McLennan's [*Lt. John C.E. McLennan*] plane (old 18 – the one that Gundy & I had the fight in) begin to waver and I expected it to go down. The rear of our formation began to creep up on us and I knew that things were getting hot. The Bosche kept after us until we got to Lamorville where they turned back. By the time we crossed the lines three of our planes were ahead of the leader. They kept sinking and I was afraid that some of them would crash before they got home. I had to watch 2 planes that were riding about 1000 meters higher than we were as I was afraid that they were Bosche but evidently they were an allied patrol.

When we got home we learned that Lt. Alexander [*Lt. Arthur Alexander*] had been shot through the back – McLennan [*Lt. John C.E. McLennan*] was shot in both legs. Lt. Don Warner [*Lt. Donald D. Warner*] was shot in the thigh and his leg was shattered. Lt. Hexter [*Lt. Avrome Hexter*] had two bullet wounds above the right eye.

We certainly were lucky to get back. Alexander sure showed "guts" to bring his machine back being that he was shot through the back. Major has recommended Alex and Don Warner for citations. They certainly deserve them. Were going to have another raid but didn't have enough planes ready.

We have meeting and decide that all Bosche planes brought down will be credited to the squadron unless it is a clear cut case of one man bringing him down. In evening Major Dunsworth takes Chick Evans and I over to see Swede [*Lt. Cecil G. Sellers*] and Hooper [*Lt. Thornton D. Hooper*]. They are coming over here [*Amanty*] as soon as they get outfitted.

September 5 – Thursday

Rain this morning. Putter around in morning and afternoon. In evening go to Neuf [*Neufchateau*] with major and Gundelach. Go out to hospital to see Alex, Warner & McLennan. They are all getting along quite well. Have dinner at L -- _ Club. See Johnny Cooper & Penn. Bring Tom Farnsworth & Bates [*Lt. Edmond Bates*] home with us. Pass some ambulances and see Jim [*blank space for last name*] from Los Angeles on one of them.

September 6 – Friday

Rain today. Stand by all day but no raid. Hooper and Swede's [*Cecil "Swede" Sellers'*] squadrons start to move over [*Thornton Hooper was commanding officer of 11th Aero Squadron and Cecil "Swede" Sellers was commanding officer of 20th Aero Squadron. Both of these squadrons flew D.H. 4 aircraft and would operate from Amanty airfield*]. [*illegible*] three machines. We get 4 new trans [*personnel transfers*]. Gundy and I have a set to about "A" and "B" flights. He is a hard man to talk to. Have a hard time making out the flight substitutions.

September 7 – Saturday

Major Dunsworth is made Group Commander of the 1st Day Bombardment [*First Day Bombardment Group*]. He makes me Group Operation Officer. Was a surprise to me. He makes Pinky Way Squadron Operation Officer [*96th Squadron*] in my place. Gundelach is made Flying Officer. [*Captain James A. Summersett, Jr., 1891-1974, was made commanding officer of the 96th.*] Other squadron operation officers report to me. Keeps me busy making out reports and orders.

September 8 – 9 – 10

Not much doing. Rains continually not possible to bomb. Gundy, Chick, Gaylord, Bates, Codman & Hexter & I go to N [*Nancy*]. Have a great dinner. Gundy buys wall paper for our new room. Hexter sticks right with us all day. Frenchman plays piano and Charley [*Codman*] does shadow dancing. Hexter appears on the scenes. Charley goes to sleep. Normid [*?*] the cook comes in and entertains us. Speaks of Lt. Codman as Charles. Cecil Young salutes him. Cold ride home. Get home at 2 a.m.

We are finally getting settled in our new quarters. The old roof leaks and it keeps Chick & Gundy busy moving their beds. I really have got the best place although I expected to get the poorest place since I lost in the draw. It doesn't rain in my place.

September 11
Still raining today not much to do. In afternoon Major D. [*Major Dunsworth*] goes over to Hdq. Gets back about 7 o'clock. Calls a meeting of squadron commanders. Also myself. Says attack [*St. Mihiel Offensive*] is to start at 8 o'clock on the morning of the 12th. Four teams from this group are to report in the morning to Hdq. to act as special couriers. Major asks me who to send. I say as observers – Coryell [*Lt. Ralph I. Coryell*] & R. E. Thompson of the 96th. He agrees. I tell Coryell to get ready – go to barracks Pinky Way, Gundy and Charley Codman. Tell Pinky about observers. He doesn't like it because he didn't know anything about it. So I go back to the Major and suggest that Penn pick the men to go. He suggests Pressler & O'Toole. Major finally agrees on O'Toole [*Lt. James A. O'Toole*] & Stuart [*Lt. William A. Stuart*]. Every body is to stand by at day break for missions.

Chapter 11

The St. Mihiel Offensive
September 12-16, 1918

September 12

    The offensive is on. There was a terrible bombardment last night. We all stand by at eight o'clock. We get orders to bomb Buxieres. It is rotten weather and the clouds are very low. Gundy says it is impossible to fly a formation in these clouds so he says one machine should go over alone. Pinky Way says well as long as you are chief flying officer & I am chief observer we will go first.

    I watch them go off of the ground for I have a hunch something is going to happen. Pinky wears the teddy bear with the devil on it. They left at 10:45 for Buxieres. The wind is pretty bad and conditions aren't very good. They should have come back in an hour but at noon they are still missing. D. Young [*David H. Young*] & Sam Lunt lead a formation of eight planes at 1:30 to bomb Buxieres. They go in bad weather, clouds very low, at 800 meters. Their bombs drop south of Buxieres. No E.A. seen and they all get back safely.

    We get orders to send a formation out again. Gaylord and I try twice to get off of the ground but we broke our propeller in the mud twice. Last time I almost had to go with Capt. Summersett but Hexter went instead. It got raining and very dark and no machines turn up. We begin to worry. Get Beverley [*Lt. Belmont F. Beverley*] the flare officer out and get the flares going. At 8:30 we hear the machines coming and will never forget how one looked as it flashed in the flare and hung there. A little later there was a terrific crash and we sure thought some one was killed. L. Turnbull [*Lt. Lewis F. Turnbull*] & Cawston [*Lt. Arthur H. Cawston*] land in tree but get out without a scratch. Captain Summersett got down alright but some one crashes into him. Bates [*Lt. Edmond Bates*], I think. Cronin [*Lt. Edward M. Cronin*] & Bleeker [*Lt. Lyman C. Bleeker*] have a forced landing at Gondrecourt & Cronin is killed – Bleeker unhurt.

It was almost dark when they bombed but they hit the road between Ugneville [?] & Hallonville [?]. Some hospital telephones in evening that Lt. Gundelach body was found near Commercy but wanted to know if he had an observer. Later they telephone that Penn Way had been found burned in the machine. This is a terrific blow to all of us. Later we hear that Gundy & Penn's machine was seen to join up with a squadron and shortly afterwards dove down in flames. Gundy was seen to jump from the plane but Penn stayed in and was burned.

How vacant our room seems with Gundy & Penn gone. It's pretty lonesome for Chick [*Elisha E. Evans*], Charley Codman and me.

Capt. Summersett was made C.O. of the 96 today.[37] He makes Chick Evans operation officer.

Pat Anderson, Hugh Thompson, Stuart [*Lt. William A. Stuart*] and Anspach [*Lt. Ralph Anspach*] overstay their leave at Nancy. Don't come in tonight.

There was more drama to the first mission of September 12th than Howard Rath confided to his diary. The weather, as relates the 96th Squadron history, "was on all counts the worst flying day in many months. A terrific southwest wind made formation flying extremely dangerous, and the low fast-moving clouds made it impossible to see more than two or three kilometers [1.2 to 1.8 miles]."[38] Pilot Charles Codman, a witness, wrote that flight leader Andre Gundelach remarked that they would not be able to fly until the weather broke, adding that to fly in the bad weather would be to waste airplanes and men. When Gundelach was told that he was to fly the mission as ordered he stalked into the Operations Building to see the Commanding Officer and closed the door behind him. When

---

[37] Captain James. A. Summersett was 27 years old.
[38] Gorrell, Series E, vol. 14, p. 16.

FIRST TO BOMB

Gundelach came out he spoke to Pennington Way, his observer, who then climbed into the rear cockpit. Gundelach told the pilots and observers to wait for him and Way to return. Way, who had recently written to his wife to say that so far he had lived "a charmed life," grinned and waved a hand as they slowly moved across the flying field, which was a sea of mud. After a long take-off roll the plane lumbered into the sky and disappeared into the thick weather.[39]

The final mission of the day also carried drama and tragedy that would endure long after the war. The scheduled take-off was 5 p.m., but around 4 p.m. four Salmson observation planes, lost and low on fuel, came in for a landing. The wind was still blowing fiercely. One pilot made a poor landing crashing into two of the eight Breguets lined up in front of the hangars while being readied for the mission. At 6:35 p.m. five Breguets managed to get into the air. The sun was going down; they would be forced to bomb and return in the dark.[40]

In 1925, Howard Rath testified for the prosecution in the court-martial of former Brigadier General William "Billy" Mitchell, who on September 12, 1918 was Chief of Air Service, First Army. Rath testified that the commander of the First Day Bombardment Group, Major James L. Dunsworth, protested the order to fly the mission when it was obvious that not only would these day bombardment flyers be over the target in the dark and thus unable to see the target, making the chance of the raid being successful extremely remote, but they would be forced to return in the dark. They were not trained to fly at night, and the field was not lighted, making the risk to men and equipment very high. As it happened, Billy Mitchell was at the airfield on the late afternoon of September 12th. Mitchell was awaiting replacement of the propellor of his aircraft which had been broken by mud during a take-off attempt. Howard Rath testified that Major Dunsworth, after his conversation with General Mitchell protesting the mission, told Rath that Mitchell had said, "yes you can make it. They should go."[41]

---

[39] Codman, Charles R., *Contact*, Boston: Little, Brown and Company, 1937, p. 81-81. *Pennsylvania: A Record of the University's Men in the Great War* (n.d. 1920) p. 36 for "charmed life" quote.

[40] Hudson, *Hostile Skies*, p. 156.

[41] Mitchell, William, *Court Martial Trial Transcript*. NARA M1140. Secret and Confidential Correspondence of the Office of the Chief of Naval Operations and

Figure 10. German photo of crashed unidentified 96th Sq. aircraft.

September 13 – Friday

Weather still very bad. Lt. Brad Gaylord & I are to lead the first bombing patrol across but the weather is so bad that we give it up until at 3:35 p.m. (4:45 p.m.) we are ordered out to bomb any concentration around Chambley. Brad Gaylord & I lead the flight. Tom Farnsworth & R.E. Thompson flying deputy, 2nd Lt. S. Hopkins [*Lt. Stephen Hopkins*] and 1st Lt. B. Williams [*Bertram Williams*], Lt N.C. Rogers [*Newton C. Rogers*] & K. P. Strawn [*Lt. Kenneth P. Strawn*], C.P. Anderson [*Lt. Charles "Pat" Anderson*] & Lt McDowell [*Lt. Stewart A. McDowell*]. Only Farnsworth, Hopkins and our machine were the only ones that got up in the air on account of the mud.

We had to keep fairly low on account of the clouds. Had a tight formation as we got over Toul. Could see towns burning as we approached the lines. We start up on east side of R.R. but can see no troop movements. So we turn to the N.W. to bomb Chambley. Just as we turn the Germans shot some shrapnel and we can see some bosche planes diving through the shrapnel at us. They are about 1000 meters higher and just as we drop our bombs they get on us. A fire starts in the town. Looks as if we had hit a dump of some kind with

---

the Office of the Secretary of the Navy, 1919-1927. p. 2967, 3002, 3003. Available online at fold3.com.

FIRST TO BOMB

our bombs but had no time to look further as the 15 Bosche planes are right in among us.

Just as we had turned towards Chambley to bomb Tom Farnsworth shot by us on the left and Hopkins banked on the right. The Bosche - about five of them landed on Hopkins and in a moment he was going down in a steep dive with about 4 on his tail. Tom Farnsworth was catching up with us and we turned to pick him up. All the time I was shooting bursts at the Bosche and I could see Robert Thompsons guns throwing out a sheet of flame. Just before Hopkins went down one of the Bosche slid right down in fact never came out of his first dive.

Just after we had turned back to the lines Tom Farnsworth's machine went down in a steep glide. Rbt. Thompson still shooting. That left our machine alone and I thought my last minute had come as there were at least 3 Bosche riding right behind us. Brad Gaylord maneuvered a good deal and finally dove in a cloud just as the 3 Bosche were closing in. When came out of the other side of the cloud the Bosche had lost some distance but tried to regain it and 3 more Bosche coming back from a patrol pulled up at us and took a few shots but the worst was over. As we crossed the lines 3 of our chasse dove across the lines and the Bosche beat it back.

What a relief it was to get across and yet it didn't seem right to come back without the others. We could see the American artillery whaling away from the woods and could see the [*troop*] concentrations on the roads. Passed right over Toul and that was a big relief to get that much behind us. Was trying to fix my gun tourelle all the time as it wouldn't raise and I was afraid some Bosche would pique on us any moment.

We got back home at 16:40 and it was terrible. Chick Evans came walking out and said "where are the rest of them" and Brad said "They're gone." The camp certainly was blue. All other operations were off although there was supposed to have been another. There wasn't any hole in our plane that I know of. Evidently the fellow who was shooting seemingly right at us never got on us. I will never forget that sheet of flame that was coming out of his guns – it looked about four feet broad and thin like a blade of fire. Any moment I expected it to cut us in to [*two*] and I could only wonder how it would feel. I thought, well Pinky Way went down one way and here we go in another way.

Poor Pat Anderson when he heard about Tom Farnsworth was terribly low but he couldn't be lower than Brad and I were. Charley Codman walks around like a man in a dream – he certainly has the appearance of a man who thinks he is going to be killed. The new fellows that just came into the squadron certainly have got their wind up and I don't blame them.

Rath, as well as the others involved in this combat, would receive the D.S.C.

The losses suffered by the 96th squadron were devastating to the morale of the airmen as Rath records in his diary. Incredibly however, despite the deaths of eight airmen, 25% of the squadron, 96th pilot Bruce C. Hopper was to write in the official history of the squadron immediately after the war that "The losses of the first two days in no way disheartened the fliers of the 96th." [42] [43] *Charles Codman sarcastically wrote in reaction to that statement, "Well, perhaps not. However, the lack of planes could hardly be termed heartening and with increasing motor trouble, the number of machines actually able to reach the objectives was as a rule pitifully small."*[44]

In his testimony, in 1925, before the U.S. House of Representatives Select Committee into Air Service Operations, Major Thomas DeWitt Milling, who in October 1918 succeeded Billy Mitchell as Chief of the Air Service, First Army said, "During the Battle of St. Mihiel we had terrific losses in our day bombardment." And *"our losses were so great at the beginning of the battle that the morale of the group itself was very low. In fact, we did not know whether we would be even able to continue, the losses were so great."*[45]

---

[42] Gorrell, Series E, Volume 14, p. 18, 50-51.

[43] Harrington, Hugh T., "'We Were Told What to Write,' a Closer Look at Gorrell's History," *Over the Front*, The League of World War I Aviation Historians, vol. 36, no. 1, Spring 2021, p. 4-17.

[44] Codman, Charles R., *Contact*, Boston: Little, Brown and Company, 1937. p. 91.

[45] *Inquiry into Operations of the United States Air Services, Hearing Before the Select Committee of Inquiry into Operations of the United States Air Service,*

FIRST TO BOMB

September 14 – Saturday

Flight of the 96th [*squadron*] left at 6:25 to bomb Conflans. Dave Young & Sam Lunt lead the formation. Weather was still bad but it had to be done. They made good hits and saw no Bosche. They were at 4500 meters and got back at 8:20. Only 3 planes got to Conflans, Dave Young, Sam Lunt, Pat Anderson & Hugh Thompson, Hopper [*Lt. Bruce C. Hopper*] & Kelly [*Lt. Arthur Kelly*]. They all got back safely and everyone feels better.

The 20 [*squadron*] leaves field at 7:15 a.m. to bomb Conflans. Sellers & Payne [*Karl C. Payne*] lead. 8 planes get to objective. Their bombs hit the town of Conflans. They all get back. 9 E.A. were seen – 6 tried to catch them but couldn't and the other 3 didn't try to. They report considerable activity on R.R. south of Conflans.

The 11th squadron leaves the field at 6:45 a.m. – Chapin [*Roger F. Chapin*] & Laird [*Clair B. Laird*] lead. 7 reach objective. They make a good hit in center of main yards and in among warehouses but are attacked by 9 Bosche right afterwards. 2 of their planes were seen to go down and one E.A. This happened between Conflans and Puxieux. E.A. had red noses & tails – probably a circus of Fokkers.

At 11 o'clock the 96th went out to bomb Conflans but it was so cloudy they bombed Etain. Gaylord [*Lt. Bradley J. Gaylord*] & Cawston [Lt. Arthur *J. Cawson*] had to lead as Beverley [*Lt. Belmont Beverley*] & Newbury [*Lt. Frank J. Newbury*] dropped out. They bombed Etain and hit the middle of the town – altitude 4500 meters. Saw no Bosche and all got back OK.

The 11th [*squadron*] started out about the same time but came back with their bombs on account of clouds. Altitude 10,000 feet. No Bosche seen.

The 20th [*squadron*] bombed Dommary Baroncourt on their second trip. They couldn't see their bursts on acct of clouds. No E.A. and they all got back OK.

At 1:15 the 96th left to bomb Vittonville & Arnaville or any concentration on the roads. D. Young & Sam Lunt led. Weather still

---

House of Representatives, Sixty Eighth Congress, on Matters Relating to the Operation of the United States Air Services, Washington DC, Government Printing Office, 1925, p. 2269-2270.

bad. They only got an altitude of 3800 meters. Raid turned out very good. They had no trouble and hit Vittonville & Arnaville & among troops in a field.

The 20th went out to bomb Bayonville at 11:05. Sellers and Parrott [*Lt. E.A. Parrott*] lead – 10 Liberty DH leave field – 4 reach objective. They cut road at Bayonville. As the 20th got off the ground one of their planes crashed & pilot Lt. C. G. Stephens was killed Lt J. J. Harris seriously wounded. The rest got back across the lines altho one had its engine fly to pieces in the air.

At 3:30 the 20th went out again for bombing Gorge, Sellers and Parrott lead. 10 started – 3 reached objective. Bombs burst in town and to east of town. They were attacked by 4 Fokkers (red noses & white tails) between Cont-a-Mousson & Gorge but without result. One enemy plane seen in distance between Nancy & Toul.

Today has been a good day as far as casualties concerned.

September 15 – Sunday

Weather still rotten not good enough for flying but still we have to go to work. At eleven o'clock we get an order to go out on a special mission to bomb Arnaville. 7 left 6 reached objective. They left at 11 and returned at 13 [*1:00 pm*] o'clock. 9 E.A. attacked over Arnaville but not very vigorously. Greenish brown Fokkers. Some think they are Austrian Fokkers. The Bombing was good. Photo shows bursts & incendiaries right in the town. All got back safely.

The 20th went out at 11:05 - 10 planes left – only four reached objective – Sellers & Parrott led. Hooper & Strauch [*Lt. Harry H. Strauch*] led a formation of the 11th at the same time. 9 of the 11th left but only two reached objective. Hoop's bunch joined up with Sellers and Hoop claimed that Sellers went all over Germany. Swede [*Sellers*] claims he bombed Bayonville – Hooper says R.R. between Onaville [?] & Arnaville. They run into the same E.A. - nine of them but nothing much happens.

Everybody is tired out with getting only about 4 hours sleep at night and working all day.

Second mission left at 15:15 o'clock to bomb Longuyon. In the 96th Gaylord [*Lt. Bradley J. Gaylord*] & Lunt [*Lt. Samuel M. Lunt*] lead. 7 planes left but only 4 reached objective. The 20th went out the same time – 6 planes left but only 1 [*originally "4" but overwritten*

FIRST TO BOMB

with "1"] reached objective – Lt. Sellers & Parrott. They joined on the 96th formation & flew with them. Bursts were fair. They hit eastern end of yards & among buildings. 6 planes followed them back from Longuyon but did not come close. Was a good concert at the Y tonight. Looks like a raid tomorrow.

Sept 16 – Monday

Weather just passable this morning. Dave Young and I lead formation [*of the 96th*]. 7 planes left field and 7 crossed lines. Our motor went bad at Bar-le-duc but as it picked up again we led across. The 11th left right after us and was to come along with us but they [*illegible*] held off at a distance. Every thing was going along fine – and I had taken my sight and gotten my bombing time and we were just coming up on the objective when our engine went bad again and Dave told me to shoot the signal that we were going to drop out and go back. This I did but all of the other planes turned with us and when I dropped my bombs all of them dropped theirs too – instead of carrying on with the bombing raid and going over the objective.

I had visions of going down and being a prisoner in Germany as we had to go back against a stiff wind and Longuyon is almost due North of our lines. We were in luck because there was no E.A. and we could take our time about getting back.

As we turned back, I could see the 11th over Dommary don't know how they got way over there. As usual in the early morning the visibility was great and don't think I ever got so much reconnaissance as I did this morning.

Expected to see about half a dozen Bosche pile on us but we didn't have any trouble at all. We recrossed the lines at 3000 meters and tried to make our field but had to come down at Belrain. Another Breguet #5 came down with us as it had a hot motor. We came down on a French field and I got the major on the phone. I turned my reconnaissance in and then Dave Young and I got in the other machine and came home. Had to send mechanics up to fix the other machine and it didn't get back until the next day.

It seems that only 2 of the 20th got across and they bombed Longuyon. About 5 of the 11th got to the objective out of nine. They say they bombed Longuyon but I think that they bombed Dommary. They all came back across the lines. They say their bombs hit near the

FIRST TO BOMB

barracks. At 12:35 Gaylord & Sam Lunt led a formation of 8 machines to Conflans but only 6 got across the lines. Altitude 4200. They ran into no Bosche then missed the target, the bombs dropping north & east but got good observations. Lots of movement north. Everybody got home safely.

The 20th went over about the same time to bomb Conflans. 8 left but only 5 reached Conflans. Went over at 12000 feet. They hit the R.R. again, they claim to have got a few bursts on the tracks. 7 E.A. red noses & wings and white fuselages followed from Conflans to Chambley but did not attack. Not much observation.

The 11th didn't go over the lines even to drop their bombs. Brought them back home. The major sure was sore.

The 96th was called out to bomb Conflans again at 16:50 [4:50 pm] o'clock. Seven left the field but only 5 crossed the lines. Elliott [*Lt. R.P. "Bob" Elliott*] and Ellis [*Lt. Arthur C. Ellis*] couldn't keep up so they dropped their bombs across the lines near Vigneulles and came home. They say the other four machines were in good formation at that time.

The eleventh squadron was started off just for this flight, the 96th second, and the 20th 3rd. Each flight was to leave the field on signal from me (a six green ball flare). But for some reason the 11th started off 2 minutes ahead of time. Their leader said his adjutant said to leave ground when flare was fired and that the six green wasn't specified. The 96th of course wasn't ready so there was about 5 minutes difference between the time they left the ground. The 20th left the ground right after the 96th. They all were to follow the same route and keep in sight of each other and protect each other. Only Swede Sellers & Payne got across of the 20th – they never saw the 96th or 11th. They bombed Longuyon & were chased all the way back to Vigneulles by 7 Bosche that just missed catching them. The Libertys were just a little too fast for them.

The 11th came back and reported that they circled around before they

[*The Diary ends at the bottom of the last page of the book with this incomplete sentence. Written up the edge of the page is "See old diary book for the rest in back of book." What follows is from the "old diary" with the completion of the incomplete sentence.*]

before they went across. 4 of them finally went across & bombed Conflans but their bombs fell short of the objective.

We waited for the fellows from the 96th to get back never thinking that anything had gone wrong. Chick Evans laughed when I suggested that we had better get the flares out. We kept flares out until 9 o'clock but none of them came back.

Those missing are: 1st Lt. C.P. Anderson [*Charles "Pat" Anderson*], 1st Lt. H.S. Thompson [*Hugh S. Thompson*]; 1st Lt. C.R. Codman [*Charles R. Codman*], 1st. Lt. S.A. McDowell [*Stewart A. McDowell*]; 1st Lt. R.C. Taylor [*Raymond C. Taylor*], 1st Lt. W.A. Stewart [*William A. Stuart*]; 1st Lt. N.C. Rogers [*Newton C. Rogers*], 1st Lt. K.P. Strawn [*Kenneth P. Strawn*].

This is getting to be one blow on top of another but we are still hoping to hear from them in the morning. It sure is lonesome for Chick and I in our barracks. Charley Codman[46] leaves us all alone now. Gundelach & Pinky Way went West on the 12th.

Sept 17 Tuesday

The weather was worse than ever today – but we got up at 4 o'clock as usual. I am beginning to feel the loss of sleep. We wait for orders but thank the Lord none come in. The Americans seem to have taken all they want to, just at present – the St. Mihiel "hernia" has been completely wiped out. No news from our four teams and every body is terribly blue. Poor Pat and Charley Codman – they were both friends.

---

[46] *Charles R. Codman graduated from Harvard in 1915, after which he served a year in the Massachusetts Volunteer Militia. For nine months he served in France with the American Field Service as a driver. In April 1917, he passed through ground school at M.I.T. and the primary flying school at Essington, Pennsylvania, receiving his commission as 1st Lieutenant in October 1917. He was sent for advanced flight training at Issoudun. From Issoudun Codman went to the 7th Aviation Instruction Center at Clermont-Ferrand arriving in early February 1918.*

Codman was 25 years old when he failed to return from the raid on Conflans. Unknown to Howard Rath, Codman and his observer Stewart McDowell were not killed. After being shot down they were captured and sat out the war in a POW camp. Codman returned to service in WWII where he became an aide to General George C. Patton.

FIRST TO BOMB

The major and I take a physical examination the doc says it means promotion – I don't care much. I am so blue. We hang around all day waiting for orders but none came. It would just been another case of big loses to go out in this weather. Still no news of our fellows.

Colonel Rader [*Ira. A. Rader*] blows into camp – he talks about his "staff." Doc [*possibly Doc Lee*], LeRoy [*Lt. C. H. LeRoy, Intelligence officer*], Chick and I go to Badonvilliers [*Badonvilliers-Gerauvilliers*] for dinner – to forget our troubles. Have a duck [?] dinner. The woman up there is very blue because Pat [*Charles "Pat" Anderson*] and Tom [*Thomas Farnsworth*] are gone. Avrome Hexter is up there and talks as usual. Have a funny walk home. Chick and I are sent for as soon as we get home. Major gets orders to have raids started in the morning. Seems to be concentrations at Mars-la-Tour.

September 18 – Wednesday

It was impossible to get off all day on account of the weather and finally Major D. [*Major Dunsworth*] called every thing off and I went up to the barracks to change my clothes. But I had no sooner got up there when [*illegible – Lt. C.?*] ordered us out and the 11th & 20th got ready to go.

The 20th left at 4:24 p.m. and the 11th at 4:45 p.m. Hooper had sent a number of his pilots that were to go on this raid over to Colombey to ferry ships as he thought the raid was off, so he went on this raid himself. Of the 11th, 10 planes left the ground – 6 crossed the lines. Of the 20th – 7 left the field and 5 reached the lines.

The 20th got up to 10000 ft but on acct [*account*] of the clouds dropped their bombs near Abbeville. They couldn't see what they hit. They saw a fight going on about 2000 ft below them and saw two planes go down in flames. These were undoubtedly of the 11th. The 20th thought it was a fight between Spads & E.A. and as the odds were so much in favor of the enemy they didn't go down to help them. They saw about 7 Fokkers which were black with white stripes. All of the 20th planes came back.

Lt. Oatis [*Lt. Vincent P. Oatis*] and Guthrie [*Lt. Ramon Guthrie*] of the 11th came in at 6:30 and said "we are the formation." They said they were attacked at Lachaussee just as they were coming through a cloud. It seems they had a good formation until they hit this cloud and that when they came out their formation was all broken up and just then 11 Bosche pounced on them out of the clouds. Three of our planes were seen to go down in flames and one bosche. The E.A. had white bodies with white stripes – up and down to cockpit and then horizontal to tail.

Oatis & Guthrie had had cloud flying so they ducked in the clouds and flew south by compass. One Bosche attacked them about 10 kilo back of our lines. They claim they were followed attacked all the way to this place but just as the first attacked they shot down one E.A.

We hoped to hear from the other fellows any moment but none comes. Just think of little (corporal) Hooper gone – my first pilot – one of the whitest men that ever lived. He would never ask any one to do what he wouldn't do himself. Just that morning he had come to me and said that "Gee here I have to come to a 2nd Lt. for orders." That part was a joke of course. Hooper just the night before had said "Rath as a pilot speaking to his old observer what do you think of some of these things."

Those missing are:
- 1st Lt. T. Hooper [*Thornton D. Hooper*]
- 1st Lt. R. R. Root [*Ralph R. Root*]
- 1st Lt. Harter [*Lester S. Harter*]
- 1st Lt. M. Stephenson [*McRea Stephenson*]
- 1st Lt. J.C. Tyler [*John C. Tyler*]
- 1st Lt. H. H. Strauch [*Harry H. Strauch*]
- 1st Lt. R. F. Chapin [*Roger F. Chapin*]
- 2nd Lt. C. B. Laird [*Clair B. Laird*]
- 1st Lt. E. T. Comegys [*Edward T. Comegys*]
- 2nd Lt. A. S. Carter [*Arthur S. Carter*]

The plane that came back still had its bombs on and they say they suppose all of them had to go through the fight with them on acct of the bomb releases not working. This has happened time and time again with the [*illegible*] DH 4s.

September 19th Thursday

Bad weather – nothing doing – although all got up at 5 o'clock. Chick and I move over to Doc Lee's barracks. It is too lonesome where we are. The new pilots come in and sure have got the wind up on account of our losses. Poor movies – lots of ragging goes on.

FIRST TO BOMB

September 20th Friday

Weather still bad. Some practice flights by the 11th. More machines come in. The 96th is pretty well filled up now. Have fun ragging Steve. Pressler [*Lt. Warren S. Pressler*] and C.P. Young [*Lt. Cecil P. Young*] quit flying as well as Cawston[47] [*Lt. Arthur H. Cawston*] & Anspach[48] [*Lt. Ralph Anspach*]. Capronis are out tonight. Sit around stove and listen to Major D. [*Major James Dunsworth*] & Colonel Rader [*Colonel Ira Rader*] swap stories about flying in Mexico. Steve [*Roger G. Stephens*] comes back from Nancy at 2 o'clock A.M. Gets cold tonight – winter is coming.

Rath wrote in his "Origin and Development of American Aerial Day Bombing During World War I" that "After operating for several weeks in this extremely bad weather, the courier, who nightly arrived at about 2:00 A.M. from Headquarters with the orders for the day, handed me a very disturbing order. The order stated that from that day on we would only report the total weight of bombs dropped <u>across the lines</u>, not, as formerly, how many bombs we had dropped on military bombing objectives. We who prided ourselves on our precision bombing were very discouraged by this change. As the reason for such an order, anyone is free to draw his own conclusion."[49]

September 21 – Saturday

Got up early to stand by but it was bad all day. A few letters from home. Lt. Brainard [*Lt. Spencer Brainard*] comes into camp to visit. Think we lack good leaders. He is over with French and they only went out once the first day of the drive. We play a joke on Steve [*Robert G. Stephens*, supply officer] and have Doc Lee [*Lt. William A. Lee, M.C.*] examine him as to going on flying duty. Sure gets

---

[47] 1st Lt. Arthur H. Cawston, observer, was assigned to the 7th A.I.C. staff as an instructor in aerial gunnery. See *Flights and Landings* (7th A.I.C. newspapers), Dec. 11, 1918, p. 2.

[48] 1st Lt. Ralph Anspach, observer, was assigned to the Information Section on October 25, 1918, relieved on November 30, 1918. See Gorrell, Series L, Vol. 4., p. 19.

[49] Rath, Howard G., "Origin and Development of American Aerial Day Bombing During World War I," unpublished typescript, p. 23.

Steve's wind up. He swore he "has got good eyes" and isn't going to look for anything but Bosche.

September 22 – Sunday

Rain again today. Still on the alert. Swede Sellers was up at a hospital and got Penn's [*Pennington Way*] watch, his wife picture and his tag. I wrote to Natalie Lucas about his death today.[50] We get orders to move today. We go out tomorrow. We are all low today.

Pennington Way was married to Eleanor Gosling (1892-1988) and they had two children. Rath wrote in the letter to Natalie Lucas that "I know it will be some comfort to Mrs. Pennington Way to know that a photograph of her and the children was found in his pocket..." Lt. Pennington Way was 28 years old.

September 23 – Monday

We all get busy moving. Major & LeRoy leave early in the morning. A rotten looking day. After lunch Col. Rader, Doc Lee, Steve and myself get in a Dodge and start out to new field. We get to field only to find that our location has been changed. We tour the aviation fields of France finally decide to go back to B-C [?] for dinner. We run into Hexter and he tells us about new location. We get to a chateau at about nine at night. Dogs bark, girl comes to the iron gate – Steve says he doesn't care to go further. Finally get out to field and such a mess. One private expresses my opinion when he says "I wonder if they think we are damn gypsies." The Col., Doc Lee & Steve, [*illegible*] and I sleep in the hangar. Rain drips on me – makes me think of a circus moving.

September 24 – Tuesday [*The First Day Bombardment Group's new location is at Maulan, 18 miles NW of Amanty*]

Had a good sleep. Everybody getting located. Chick shows up from B [?]. It's a hell of a field. Our table and eats are under the trees.

---

[50] The letter to Natalie Lucas, a good friend of Pennington Way's wife, is found in *Descendants of Robert and Hannah Hickman Way of Chester County*, by D.H. Way, privately printed, 1975, p. 62. Howard Rath is misidentified as "Roth."

Everybody is sore, because they have to sleep out in camp tonight. Thought we had to go to work tomorrow but evidently not.

September 25 – Wednesday
Cold last night – winter is coming. Chick and I go to Ligny [*Ligny-en-Barrois, 3.5 miles east of Maulan*] to get a bath – but there is no water until afternoon so we are S.O.L.

Putter around in afternoon trying to get squadron operation officers lined up for the big show. Get some letters from home – that helps some. In evening sit around in my room – Chick, the major, LeRoy. Major and I talk after rest go. Looks like Hex [*Avrome Hexter*] will get a medal. There are going to be some more squadron commanders – we poor observers are S.O.L. as usual. Big show tomorrow.

Chapter 12

Meuse-Argonne Offensive

September 26 – Thursday

Drive [*Meuse-Argonne Offensive*] started at 7:30 this morning. Barrage began at 2:30 a.m. and what a roar there was last night. It was more of a gigantic pulsation or throb than a roar. Courier came in at 2:30 with the orders and woke me up. We went on the alert at 7:15. At about 9:30 we got an order to bomb Dunn-Sur-Muse. The 96th, 11th & the 20th sent flights out. They were to meet over the [*illegible*] but never connected. The 96th bombed & were attacked just as the[y] turned. They got back safely with the exception of Lt. Paul O'Donnell (Forshay's [*Lt. Harold J. Forshay*] observer) who got hit in the legs with two bullets and bled to death on the way home. Forshay was very near crazy when he got back. The first flight of the 20th came along and protected the 96th in the fight - they got back alright too but the second flight got badly shot up – only 3 planes out of eight returned – and observer Parrott [*1st Lt. E.A. Parrott*] had been killed in one of those (Lt. Howard's [*Lt. Frank J. Howard*] machine). Poor Parrott – he was a prince.

The 11th only got one plane over and he flew with the 20th on acct of Parrott falling on the D.S. [*emergency control*] and jamming it! These 3 surviving planes toured Germany with the Bosches shooting at them constantly. Norris claims a Bosche. O'Donnell got a Bosche after he had been hit and had sunk to his knees. The 96th gets one other & the 20th - 2.

FIRST TO BOMB

In the afternoon we got orders to bomb Grand Pre [*Grandpre, 25 miles NW of Verdun*] but after part of the 96th had left the ground the orders were changed and they were to go to Etain. They were to be protected by pursuit planes but they only saw a few.

The 11th & 20th bombed Etain – no casualties. The Americans gained about 2 ½ miles on a 40 mile front N[*North*] of Verdun.

September 27 – Friday

Dud day. Stood on alert all day – but rain & clouds. Just at 4:45 p.m. when everything was called off, H.Q. orders a raid out. Major [*James L. Dunsworth*] says can't be done but we have to put one on any way. 20 Breguets get off – 12 reach lines (objective and all get back). So cloudy they only went to Etain – didn't see bursts or anything else. See some of our chasse & also a fight over Thiaucourt. No mail today. I am going over on a raid leading tomorrow with D.H. Young.

September 28 – Saturday

We started out at 7:30 to bomb Bantheville. 20 planes started but only 6 made formation. It was a wonder and the largest one I have been in. Beverley came piquing [*to dive vertically downward*] in the formation and almost crashed into D.H. Young and myself.

We could only get to 3000 meters on account of clouds. Got as far as Souilly [*23 miles north of Maulan*] when we ran into a storm – had to turn back as we couldn't see anything ahead. Could see the chasse just taking off as we passed – evidently they didn't think we were going. H.Q tried to call us off after we started.

In afternoon Colonel D [*unidentified*]. shows up and puts us through a questioning as to why we didn't bomb. Didn't call us out again – but in evening many [*?*] calls from H.Q. as to reasons. I can't get their reasoning. Major is blue – doesn't look any too good. Here goes the ball game. Decided to send two formations tomorrow – it will use every available man and that means people that aren't supposed to fly. It's enough to make any one blue. I have to go tomorrow again.

FIRST TO BOMB

September 29 – Sunday

Bad day today. G.G.I. [?] still checking up about yesterday. Worked around office all day. Wrote some letters. At 4 o'clock we get a mission to bomb Grand Pre at any altitude we can make. Clouds are very bad – doesn't look good. Chasse are to protect us. Breguets get off at 4:15. 20 leave but only 13 rendezvous. We start off at 4:30. Finally get through the clouds and up to 4000 meters. As we get up to lines can only see the ground now and then. We were going to bomb with the wind and I had my time all taken but as we got to the objective we swung around against the wind and I had to change my sight. Tried to get the big dump but couldn't keep on it so took the smaller one. Got a good burst one bomb fell in the dump – four or five on the tracks and the rest all around. One incendiary on the dump – one on the track.

The D.H. formation bombed Marq [*Marcq, 3 miles SE of Grandpre*] because they couldn't see Grand Pre ?? There was a fire burning there afterwards. Only 6 out of 20 D.H.'s reached the objective. About 14 Spads met us. It was a great trip home and we had a tight formation & the sky was beautiful. Hexter thought the Spads were Bosche. Got home before dark. G.H.Q. very much satisfied for once.

September 30 – Monday

Bad day. High wind last night & today. We stand on alert all day but make no raid – was to have bombed Grand Pre. In afternoon Major D. [*Major Dunsworth*], Chick, Krummins [?] & I go to Ligny to take a bath but couldn't get in bath house. Have dinner at the Hotel and come back right afterwards. Met officers on road. This bum water has me on the run.

October 1 – Tuesday

Bum weather still on. In afternoon we are sent out to bomb Bantheville. Breguets get there alright and get good results but D.H.'s of the 20th & 11th get mixed up on account of losing leader and all come home. Col. Milling [*Col. Thomas DeWitt Milling, Chief of Air Service, First Army, who had recently replaced General Wm. "Billy" Mitchell*] happened to be here and he starts an interview.

FIRST TO BOMB

October 2 – Wednesday

Bomb St. Juvin [*3 miles East of Grandpre*] this a.m. That is the D.H.'s of the 11th & 20th. The Breguets got over lines late. Lost an hours time picking up there formation. Couldn't see the objective on acct of cloud so bombed Cornay [*5 miles SE of Grandpre*]. Too late for protection chasse and were attacked by bosche – 7 of them. Formation badly mixed up. Lt. Walden [*2nd Lt. James P. Walden*] got shot through finger. One team, Allsop [*1st Lt. Clifford Allsopp*] and [*blank – observer was Lt. Lawrence Ward*] were seen to go down near Apremont [*3 miles SE of Cornay*] with bosche on tail. I think landed on our side.

Major Dunsworth relieved by Major Bowen [*Maj. Thomas S. Bowen, 1885-1927, age 33, the current Operations Officer for the Air Service, First Army*] this afternoon. Just my luck again. I still am operation officer but the zest is gone out of it. There is no future to it.

October 3 – Thursday

Got to get up early in the morning now and get the reports from the courier. We were on the alert again – that is the 96th – while the other two squadrons do practice formations. Didn't get away until about eleven. Go to Dun-sur-Meuse [*17 miles NNW of Verdun*] pursuit to meet them – but they weren't there – our fellows made a good hit but got in a big fight with the bosche – about 15 planes. One of our planes didn't get back – but no one saw it go down. All the planes were badly shot up.

Had to go out again in the afternoon. Kelly [*Arthur H. Kelly*] & Hopper [*Bruce C. Hopper*] led again. D.H.'s – 15 started but all came back – one landed at Neufchateau. Breguets went over at 2,000 meters. Couldn't get any higher. Just bombed when 3 groups of bosche dived on them from the clouds. Right afterwards our Spads rained in on them. Before the Spads got there our fellows brought down two bosche. It got to be a regular dog fight with the Spads. Spads get at least eleven of them. All our planes came back.

[*Entry for October 4, 1918 – Friday is out of order in diary and appears after October 5, 1918 – Saturday.*]

FIRST TO BOMB

October 5 – Saturday

The 11th & 20th bombed St. Juvin & Aincreville [*2 miles WSW of Doulcon*]. Made fair hits no casualties. 96th didn't go out.

October 4 – Friday

96 got orders to bomb St. Georges 12:40 but Landres St. Georges [*Landres-et-Saint Georges, 1.2 miles East of St. Georges*] bombed on acct of clouds. 10 planes left but 7 reached objective. Got good hits in town. No E.A. seen but 24 Spads accompanied them.

In afternoon got orders to bomb Landres St. George 12 teams 96th & 11th went out. Hopper & Kelly led. Got good hits. 30 bosche jumped them but didn't get any of our fellows. We brought down 2 of them. Afterwards a group of Spads jumped the bosche and there was a regular dog fight. Our Spads claim they brought down 11 of them. Got back at 4:30. Bombed at 5:10. [?]

A formation of 11th & 20th squadrons started out but none reached the objective.

October 6 – Sunday

96th left at daylight (10 machines – the 11th teams didn't get over in time) 9 reached the objective – Bantheville. Young & Corryell [*Ralph I. Coryell*] led. Got fair hits in edge of town and on dump. Only one E.A. seen. Plenty of Spads.

About noon the 20th & 11th started out to bomb Doulcon [*1 mile SW of Dun-Sur-Meuse*] but only 11th reached objective. Just fair hits. Got a few letters lately. Can't understand why I can't hear from Paris. Pike probably doesn't get my letters. Doesn't seem like Sunday.

The Major B. [*Thomas S. Bowen*] picks out two pilots & 2 observers to go to the states as instructors. There is an uproar on his choice. No old fellows among them. Would like to have gone myself.[51]

Learn that Major D. [*Major Dunsworth*] recommended L. Turnbull [*Lewis F. Turnbull*] for a squadron. Beverley & Young take

---

[51] The pilots may have been Manvil Davis, who joined the squadron Aug. 5th, transferred Oct. 5th and Robin G. McCord who joined Sept. 11th and transferred Oct. 5th. The observers may have been James P. Walden who joined Sept. 19th and transferred Oct. 8th and Lyman C. Bleeker who joined Sept. 5th and transferred Oct. 8th.

examinations. Must mean promotions. Some more of Major D's good work. He wasn't appreciated.

October 7, Monday

Rainy day. No raids. D.H. Y [*David H. Young*] & I go over to see D. but he wasn't in. Major D. [*Major Dunsworth*] calls up to find out if it really is so about the names – he is surprised.

Chick, George [*probably Lt. B.W. George*] & I go to Ligny for dinner. Run into some observers who are coming out to get experience over the lines before they go home to instruct. We gave them an earfull [*sic*].

Everybody seems to know about our losses. Learn tonight that Major D. recommended Gaylord & I for the D.S.C. [*Distinguished Service Cross*] also the squadron for the D.S.M. [*Distinguished Service Medal*]. Also me for a promotion. It came back for corrections – wonder why they held it up so long. Was already approved by L.C. of A. [?]. Just the regular luck.

October 8 – Tuesday

No raid today – bad weather. Major B. [*Thomas S. Bowen*] goes to Hdq. comes back saying he is going to make every second Lt. a 1st. Going to give all old pilots & observers the D.S.C. Wants me to write the stories. I can't see just how it can be done. If everybody is recommended for everything it will queer anybody's chances. He says he will write them up. At midnight he telephones that he will send them in tomorrow. Poor adj. George [2nd *Lt. B.W. George, Adjutant*] has got a job. We are all in the same lot – Simeon [?] and all of us. This is a queer world.

October 9 – Wednesday

Busy day for all of us. Lt. George busy writing up stories for D.S.C. Gaylord & mine come back again for correction. Try to make some raids on Landres St. George. 11th got there alright but 20th bomb St. Juvin. Lots of chasse. See a chasse plane go down. H.Q. sore because we didn't bomb St. George.

FIRST TO BOMB

October 10 – Thursday

Chick & the Doc went to Paris last night. Chick left Cory [*Lt. Ralph I. Coryell*] to do his work HQ asks for 3 raids by the group today. Good weather from day break. Have hard time to get squadrons started. To bomb Milly [*Milly-devant-Dun*] & 20th & 11th bomb it? but 96 bombs Dun on account of clouds. In afternoon they all go out to bomb Villers devant Dun. All get there and make good hits. 96 take a picture shows a wonderful hit. Pancoast [*Lt. Henry L. Pancoast*] and D.H. Young lead because everyone begins to follow them – Beverley was to lead. Orders come from H.Q. for the Major not to send Capt. Summersett & Lt. Ring [*1st Lt. Thomas M. Ring*] and to pick Lt. Newberry & Chick Evans or two others. He won't let Chick go but does let Frank Newberry [*Lt. Frank J. Newbury was transferred October 20, 1918*]. Glad of that much any way. I tell him that I shouldn't object to going but he says all this about "couldn't be spared" – big promotions ahead etc. I explained to him I hadn't got anything so far. He said "it was because I hadn't been under the right man"?

Good news in the paper today. English making big advance around Cambrai. Wilson's reply is a good one. No more.

October 11 – Friday

Got up early – Cory woke me – didn't like it. Fog didn't clear all day. Chick and Doc come in from Paris. They certainly had a good time.

In afternoon Gen'l Mitchell [*William "Billy" Mitchell*] came down. Major B [*Maj. Thomas S. Bowen*] says he was well pleased and says promotions will go through. Maybe! Got Beaucoup letters tonight – great – Have to get up early in the morning again.

October 12 – Saturday

Another bad day – no flying. Major [*Bowen*] gives me a spiel about Group operations officer being a Capt. Petre ! Great news in evening. Germany agrees to Pres. Wilson Terms. French have advanced way to Aisne at Rethel.

FIRST TO BOMB

October 13 – Sunday

Everybody excited about the news. A rotten day – no flying. Swede Sellers [*Cecil G. Sellers*] & Heater [*Charles L. Heater*] now Captains. George and I make arrangements about going to the front. Loaf around all day Chick Evans very sick – taken to hospital. Major [*probably Bowen*] not feeling very well.

October 14 – Monday

Rotten day. Read all day. No war news. Papers don't think that peace will come out of Germany's message. Steve [*Lt. Robert G. Stephens*], George & I go to St. Dizier in afternoon. Have dinner with Canadian girl ambulance drivers. Kelly [*Lt. Arthur H. Kelly*] was in fine form.

October 15 – Tuesday

Rainy day. Sit around. Wilson's reply to Germany's request comes in afternoon. [*President Wilson's position was that the Allies would deal only with a democratic Germany and not the imperial state ruled by the military dictatorship.*]

October 16 – Wednesday

Rotten day. LeRoy & I take walk in afternoon. No news. Steve goes to St. Dizier.

October 17 – Thursday

Another bad day. Go to Paris for four days vacation in evening.

October 18-22

Another world. Just loaf around the walks of Paris. Paris is much gayer and the Place de la Concorde is filled with war trophies. John gets promoted to 1st Lt. Hate to leave the city.

October 26 – Saturday

Not much going on but Alexander [*Lt. Arthur H. Alexander*] & Warner [*Lt. Donald D. Warner*] get the D.S.C. Miss MacArthur shows up in YMCA camp. That certainly is a joke. I get transferred to the "A Flight" of the 648 Squadron. They set up a good mess. Weather is clearing up a good deal.

FIRST TO BOMB

October 27 – Sunday

We get off a 4 squadron raid to Briquenay [*50 miles NW of Verdun*] in afternoon. Get good results especially the 96th. I hear that I am to be disciplined for writing a letter that was censored. C'est le guerre.

October 26 – Nov 5

This has been a rather colorless period a few raids with several that stand out such as on Beaumont, Mouzon and Stenay as well as Montmedy, Nouars [?], Tailly & Briquenay.

On the raid on Montmedy poor Gatton [*Lt. Cyrus J. Gatton*] was lost and yesterday in the raid on Mouzon the 20th lost 3 ships.

The day before the 166 lost 2 ships. The Huns are getting busy again. I got notice that the Distinguished Service Cross had been awarded to me. The papers had my name Howard G. Bath. Get news that our promotions are coming through at last. Things are breaking fast now, first Bulgaria goes out of the war then Turkey and now Austria. I hear that Germany has been given 4 days to comply with terms or it means Unconditional Surrender. I hope they quit.

About a hundred German prisoners are working on the roads in camp now. They are glad to be out of it and they think that initials P.W. on their trousers is for President Wilson.

We get word through a note that is dropped by a German aviator that H. Thompson [*Lt. Hugh S. Thompson*] is dead, also Stewart [*Lt. William A. Stuart*], Taylor [*Lt. Raymond Taylor*], Straugh [*Lt. Kenneth P. Strawn*]. Thank goodness Nap Hooper is only wounded slightly as is Root [*Lt. Ralph R. Root*]. Charley Codman is a prisoner. And so it goes. Tried my luck at quails several times but haven't had any luck so far. Am going on a raid tomorrow with Hooper if it is a good day. Hope my luck is still with me as there are beaucoup Huns over there now.

November 6 – Wednesday

No raid today on acct of bad weather. Lt. LeRoy [*Lt. C.H. LeRoy, Intelligence Officer*] leaves today for 2nd army. In afternoon go quail hunting but no quails. Didn't go to Infantry officers mess with YMCA girl. In evening we hear that German committee is on way to parley

FIRST TO BOMB

with Gen'l Foch [*General Ferdinand Foch, Supreme Allied Commander*] regarding armistice. Hope that it is true. Yesterday (or rather day before) Steve [*Lt. Robert G. Stephens*] and Coryell [*Lt. Ralph I. Coryell*] leave to be adjutant & operation officer of 2nd Day Bombardment Group. Will try to go on a raid again tomorrow.

November 7 – Thursday
Another bad day. We hear that our commissions have come through. This afternoon we get word that Germany is sending a committee over the lines to interview Foch regarding an armistice. Get some letters from home.

November 8 – Friday
No chance to raid today. Weather still bad. Have a talk with German prisoners. They are tired of the war and they think that the P.W. on their trousers stands for President Wilson. The Republicans seem to have come out pretty well in the election. Word comes this afternoon that Foch has met the German committee and have given them 72 hours to give him an answer. Yes or No. If no answer at the end of that time it means war to bitter end. Rain again tonight. Some more letters.

November 9 – Saturday
No raid today. Get wireless news in regard to negotiations. Germans are doing lots of talking with their headquarters at Spa. Looks as if they will accept. Sellers, Gaylord and I get orders to appear at the 1st Pursuit Group to get decorated with the D.S.C. I do some sewing in the evening. Crimmins [*? or Krummins?*] and George get into a game. I get to bed at about 2 o'clock.

November 10 – Sunday
Very foggy – no raids today. Swede [*Cecil G. Sellers*], Gaylord [*Lt. Bradley J. Gaylord*] & I go to 1st Pursuit to be decorated. Cold ride. Ceremony takes place in the fog on their flying field. Band playing, ships and men lined up. We all have to march out as our name is called. Then we line up again and one by one they read our citations and as each is read we take our place on an advanced line. Lieut. General Liggett [*General Hunter Liggett*] of the 1st American army

walks towards us and as Col. Atkinson hands him the medals he pins them on us, congratulates us and shakes hands with us. He is a regular fellow. Says we should take care of ourselves and hopes that we get home in order to wear them there. Also adds that without the aid of the aviation the 1st Army couldn't have gained as it did. Col. Atkinson also congratulates me. Run across Mr. Driggs again. He is a fine fellow.

Don't believe I was ever colder than while standing out on the field waiting for the decorations. Could see the fellows knees shaking.

Stop in and have some tea then go on and have dinner at Bar-le-Duc.

Forgot to say that Eddie Rickenbacker & Capt. Estes were decorated at the same time.

Figure 11. Rath and Bradley J. Gaylord with their DSCs

Afterwards driving to Ruvigny to see Chick Evans [*sick in hospital*]. He certainly looks like a ghost. I didn't recognize him at first. He still has the smile left. I certainly miss him. Afterwards we drive to St. Dizier and get dinner there. Brad was going to show us a great meal but couldn't see it. Swede [*Sellers*] & I have girl sew on service chevrons. Come home in fog – chilled to the bone. We hear that Germany has accepted the terms of the armistice. They have a big celebration in camp. It looks as if the war is over.

November 11 – Monday

Everybody wants to see the medal. We get official news of the signing of the armistice. The doughboys up on the front are making good use of a last opportunity. We can hear the guns roaring all night and even in the morning. We get orders that all operations will stop at 11 o'clock. All of us are sort of lost as we don't know just what to do. In afternoon we catch a wireless in which the Germans ask the allies to have the American stop attacking around Stenay. Play cards in evening.

November 12 – Tuesday

No work today. Get news that promotions are in order. Another big bunch of recommendations for DSC & promotions going in. Get Hopper to work on them. In afternoon Col. Milling [*Col. Thomas deWitt Milling*] comes down and gives us a talk in the YMCA. Sit around in evening & read papers. Try to play cards but no luck.

November 13 – Wednesday

Beautiful day. Have some formations but wind crashes some machines. Lots of visitors today. We get two D.H. 4's with armored tank system feed and light guns also a tri-place fighting Caudron. Now that the war is over things are picking up. Paper states that Kaiser has fled to Germany and that Crown Prince has been killed by his soldiers. How the mighty have fallen. George and my recommendations go in for Captaincies.

Beaucoup letters today.

November 14 – Thursday

Great day. Worked on D.S.C. orders & took a walk after quail in afternoon. No luck.

November 15 – Friday

No news especially. There are two squadrons to go with Army of occupation into Germany. 96 & 106 probably go. Getting very wintery around here. Get some more letters from home. They seem to be anxious for us to go to Berlin. Lt. Crimmins [?] and I go to Neufchateau in Cadillac in afternoon. Wild ride. Have bath & afterwards go to Ousche [?] to see Steve & Cory. Get dinner there. Great outing there.

November 16 – Saturday

Windy day. [*illegible*] on ground. Our group has to carry messages by plane. Dave Young & I go to Dun-sur-Meuse & Souilly. Pretty bumpy. Get good view of trenches. Dun is pretty well smashed to pieces. Could see bomb craters [?]. On way back came over clouds – looks like a sea of white water underneath us. Came through the clouds right at Ligny.

Storming and was hard to make a landing. 166 team didn't get off crashed 3 planes. Turnbull [*Lt. Lewis F. Turnbull*] & Lindsay [*E.W. Lindsay*] get to Chaumont [?] [*illegible*] alright. Warner [*Lt. Donald D. Warner*] & I stage a little party before dinner. Captain Summersett invites a bunch of us to dinner. Have a funny experience in the evening.

November 17 – Sunday

Learn that only the 166 is to go to Germany. Have to write & compile a history of the group in 2 days. [*illegible*] 11th delivers messages. We buy some cloth & make A insignia. Good dinner today. Bevo [*perhaps pilot Lt. Belmont Beverley*] & Cory come over. Work on history.

November 18 – Monday

Work on history. It's a fierce job. Don't know where to begin. Hopper comes from hospital to help on history. General Mitchell comes out. 166th is to go to Germany. Get beaucoup letters. They still insist we should go to Berlin before we stop.

November 19 – Tuesday

Blue Tuesday. The History of the Group still hangs around my neck. Bruce [*Bruce C. Hopper*] he can't help as he is all tired out. Tom Reed [*Lt. Thomas R. Reed*] comes over & gets busy on it. It goes slow and it begins to look as if it won't be accomplished. Lt. Cook [?]

gives me a good steer – he rakes up an old "Daily Log" that is almost complete. We are going to finish that up and send it in. We hear that there will be no more promotions. Goodbye Captaincy. George & I debate about going to Germany with the 166th. Sit around in evening & talk to Bob Elliott, Lakin [*Paul E. Lakin*] & Bruce Hopper.

November 20 – Wednesday

166 getting ready for Germany. Col. Bowen [*Major Thomas S. Bowen was promoted*] gives up command of Group. Looks like Capt. Summersett will get it. Get History of Group finished with aid of Tom Reed of the 96th.

Stories

Goldstein stable for 2.50 no manure nigger sergeant- right dress – feet belong to nigger in back line – bring your eyes around with click.

November 21 – Thursday

Lt. George & I are going to Paris tomorrow. Get some pictures taken today with General Mitchell. Sent histories of the 20th, 11th, 166th & 96th squadrons into the 1st Pursuit Wing. Getting cold & wintry. Photo section has an explosion tonight.

[*End of Diary*]

Chapter 13

Postwar

On November 22nd, Howard Rath went to Paris, to work on the history of day bombardment. To his disgust he found that he was "told what to write."[52] He refused and asked to be transferred. He was transferred to Washington, arriving in New York on February 12, 1919, where he again was to write history. However, this time he was allowed to write the history the way he wished.

It was likely in this period that he wrote an article, "Needed – Mine Sweepers of the Air", which appeared in the publication "U.S. Air Service"[53] in April 1919. In this article he lays out his views on using pursuit aircraft to provide cover and to sweep the target area of enemy aircraft to protect the bombers. In addition, he wrote that it was essential that bombers fly in weather conditions such that they could see the ground and the target; bombers ought to be armored and that commanders should have the discretion to take weather conditions, size of formations and time of attack into consideration as "there are conditions under which day bombing results gained would not warrant the sacrifice that would be encountered."

---

[52] *Aircraft Hearings before the President's Aircraft Board*, Washington, DC: Government Printing Office, 1925, p. 1112. See also Harrington, Hugh T., "'We Were Told What to Write,' a Closer Look at Gorrell's History," *Over the Front*, The League of World War I Aviation Historians, vol. 36, no. 1, Spring 2021, p. 4-17.

[53] Rath, Howard G., Lt., "Needed – Mine Sweepers of the Air," *U.S. Air Service*, published by the Army and Navy Air Service Association, vol. 1, no. 3, April 1919, p. 18-22.

FIRST TO BOMB

He was discharged from the service on March 12, 1919. On April 29, 1919, he received his long-awaited promotion to Captain, Aviation Section, Officer Reserve Corps. His promotion to 1st Lt. on November 6th had gone unnoticed in his diary. Again a civilian, Rath returned to his investment and securities business.

On May 18, 1919, Rath appeared in a short film clip taken at De Mille Field, Hollywood California.[54] It was Air Memorial Day. Many celebrities of the day made an appearance. Aircraft were flown; speeches were made; names of the honored dead were read.

The 96th Aero Squadron returned to the United States in May 1919. Today it is known as the 96th Bomb Squadron, located at Barksdale AFB, flying B-52H Stratofortress. On the 100th anniversary of the first bombing mission targeting the railroad yard at Dommary-Baroncourt, a 96 Squadron B-52H flew over Etain, France, five miles away.

In February 1925, Secretary of War John W. Weeks was considering whether to reappoint Brig. Gen. William "Billy" Mitchell to his post of assistant Chief of the Army Air Services. Weeks determined that he would not make the decision until the U.S. House of Representatives Select Committee of Inquiry into Operations of the United States Air Service had concluded its investigation into Mitchell's many criticisms of the air service.[55]

Rath, and perhaps Brown, testified before the Select Committee. Prior to their testimony Gen. Mitchell's office stated that the testimony of these men was "an attempt to divert the committee's mind from the subject before it. The subjects they are reported to be willing to discuss - bombing operations in France during the war – is entirely irrelevant."[56]

Secretary of the Navy, Curtis D. Wilbur, was investigating as well. The Navy had also been criticized heavily by Mitchell. In the

---

[54] https://www.criticalpast.com/video/65675034840_Air-Memorial-Day_William-Fullam_Frank-A-Garbutt_H-A-Arnold

[55] Reappointment of Mitchell Deferred Until Probe Is Ended, *Evening Star* (Washington, DC), February 9, 1925, p. 1. Weeks is Puzzled, Secretary of War Ponders Over General Mitchell's Reply to Charges, *The Pomona Progress* (Pomona, California), February 9, 1925, p. 1.

[56] "Recall Mitchell on Bomb Tests" *Evening Star* (Washington, DC) February 11, 1925, p. 1.

first week of February, Howard Rath, as well as his former commanding officer, Harry M. Brown, sent a telegram to Wilbur saying in part, "In case information is desired supporting your ideas from the results to be expected from bombing under war conditions and upon Gen. Mitchell's conduct and lack of knowledge of bombing during the late war, we suggest that depositions be taken from three former officers now in Los Angeles, all of whom had actual bombing experience and two of whom were in command of bombing squadrons. We know that you are right in this controversy."[57] It is tantalizing to wonder who the third man, a squadron commander living in Los Angeles, was who joined with Rath and Brown.

Rath, but not Brown, would testify in September before the President's Aircraft Board, also known as the Morrow Board[58], which was examining the best means of developing and applying aircraft in national defense.

On December 9, 1925, Rath gave dramatic, and damaging, testimony for the prosecution in the court-martial of Colonel William "Billy" Mitchell.[59] Rath criticized Mitchell's personal command decisions by testifying, among other things, that Mitchell had ordered the First Day Bombardment group, trained for day flying, to fly a night mission with disastrous results. Rath appeared in newspapers across the country with coverage of his testimony.[60]

During World War II Captain Rath returned to service in August 1942. He was stationed at Victorville, California and Hobbs Army

---

[57] "Reappointment of Mitchell Deferred Until Probe Is Ended, Weeks Desires to Have Full Data Before Acting – Former Air Officers Offer to Tell of General's 'War Conduct," *Evening Star* (Washington, DC) February 9, 1925, p. 1.

[58] "Rath Gives Air Board Testimony, Los Angeles Bond Dealer Assails Mitchell; Is Back from Eastern Trip," *The Los Angeles Times* (Los Angeles, California), October 28, 1925, p. 14.

[59] Waller, Douglas, *A Question of Loyalty: Gen. Billy Mitchell and the Court-Martial That Gripped the Nation*, (New York: Harper Collins Publishers, 2004), p. 295-300.

[60] Bennett, James O'Donnell, "Reid Enraged by Hero Witness, Hissed by Crowd, General of Court Defends Critic of Mitchell," *Chicago Tribune* (Chicago, IL), December 10, 1925, p. 7. "Assails Mitchell on Deaths in War," *New York Times* (New York, NY), December 10, 1925, p. 10. "Mitchell Assailed for Orders Issued During World War," *The Atlanta Constitution* (Atlanta, GA), December 10, 1925, p. 1.

Airfield in Hobbs, New Mexico. Lastly, he was stationed at the Army Air Force Bombardier School at Deming, New Mexico Army Airfield as Assistant Post Operations Officer.[61] *It was during this time, or very shortly after, that he wrote his "Origin and Development of American Aerial Day bombing During World War I," an unpublished typescript in which he briefly lays out the history of the 96th Aero Squadron and the First Day Bombardment Group. He also made the case for much closer contact between the bombing operations and headquarters to help those in headquarters understand the then-new world of aerial bombardment, such as the importance of weather and friendly pursuit planes flying cover for the bombers. He also advocated larger bomber formations that could defend themselves better than small formations.*[62]

Howard Rath married and raised a family. He died May 2, 1973 in Los Angeles, California at the age of 88.[63]

---

[61] "Introducing Deming AAF Officers, Captain Howard G. Rath," *The Deming Headlight* (Deming, New Mexico), February 26, 1943, p. 1.

[62] Rath, Howard G., "Origin and Development of American Aerial Day Bombing During World War I," unpublished manuscript, circa 1940s.

[63] "Last Rites Set for Broker Howard Rath," *The Los Angeles Times* (Los Angeles, Calif.), May 3, 1973, p. 47.

Appendix

Rath's Letter Describing his first Bombing Raid

As indicated in the diary, Rath and Way were sent to the front so they could fly on bombing raids with active French squadrons. Although they were bombing instructors, neither of them had been across the lines in combat. In theory they would learn from their experiences with the French to better enable them to teach their American bombing students.

Rath wrote to his brother, Walter, a long letter dated March 27th that describes in detail his first bombing raid which took place on March 10, 1918. The letter was published by the Los Angeles Times, April 28, 1918.[64] The letter provides another view of events also mentioned in the diary:

Things have been happening fast since my last letter. First of all I was in Paris on March 8 when the Germans bombed it. We were just finishing dinner when the alarm sounded and by the time we got out in the streets everything was dark and people were taking cover. We stayed outside to get an 'eyeful' in real American style. We could hear the anti-aircraft guns booming, or rather 'woof-woofing' all around the city. Up above we could hear the humming of the French airplanes that were patrolling the sky, and as they would flash their lights on and off it seemed as if the sky was full of immense fireflies. Once when things got rather thick we saw a machine gun go ta-ta-ta-ta-ing in the air as we could see the flash of the gun from the plane.

---

[64] "Bombing the Boches, Sport for Angeleno. Former Stock Exchange Man Tells of Aerial Thrills – Rheims is Like Volcanoes on Moon," *Los Angeles Times* (Los Angeles, California), April 28, 1918, p. 21.

FIRST TO BOMB

Later on we heard a few bombs explode and saw one of the Gothas[65] falling in flames. Long afterwards when I went to bed the guns were still 'woof-woof-ing' and it was sometime before I could get to sleep.

Several of us were on our way to the front, where we hoped to get some trips across the lines as observers. Another lieutenant[66] and myself are the first two bombing (aerial) observers to be sent up, so the whole trip had a pretty keen edge on it.

Fortunately, we were sent out to a French bombing squadron that had some Americans attached to it so it made it somewhat easier for us. My bombing friend and I got out to the place about 10 o'clock at night and the captain of the squadron hunted around and finally got us billeted in a farm house. Had a good room, but we had to sleep in between feather beds. The bed was too comfortable.

The next morning, Sunday, March 10, at breakfast, the French captain said I should get ready right after the meal to go on a bombing trip. He said it in a matter of fact way just as if he were inviting me out for a walk, but somehow it didn't look just that way to me. We rode out to the field in a camion [*truck*] over a road that looked more like a row of shell holes.

When we got out on the field all the machines were turning over and getting warmed up and by the time I had crawled into my flying clothes we were already [*sic*] to go, after taking a last squint at the map, making sure that my machine guns were in good working order and that the bombs were securely on.

Finally the leader took the air and we all followed in formation, about twenty machines in all. We flew up towards a certain meeting point on our side of the line and by the time we had all arrived we were at our altitude (15000 feet) and were sure our engines were working well. We passed over the line at a pretty good clip and got a good view of everything as we went over. The line looked just like a big long yellow clay smear, the ground seems to be fairly churned up, since not so long ago the French made a slight push at this point and most of the trenches were on the French side. Rheims I could see quite

---

[65] The Gotha was a large twin engine heavy German bomber.
[66] Lt. Pennington H. Way.

plainly – that is, what is left of it. It looks like the volcanoes on the moon.

After we had gotten past the lines about ten miles, the anti-aircraft guns began to open up on us and then the fun began. At first their range was a way off and I thought the black specks were German machines, but as they began getting our range there wasn't any mistaking what they were. When the shells break they make little round clouds of pitch black smoke and as they shoot up a sort of a barrage, these black puffs lie on the air around your machine just like flocks of black ducks. Sometimes they would break below us, sometimes above us and then on the side, but we kept going right on.

After traveling for about an hour we got to our objective, which was a railroad junction and I had the pleasure of pulling the trigger for the first time and watching about 600 pounds of bombs drop down on it.

As soon as we turned back we discovered four Bosche machines trailing, but as we were too strong for them, they hung back and waited for some poor lame duck to drop out of the formation. By this time I had been standing up in the back seat for two hours and it was pretty cold in all that wind. I could feel my nose beginning to freeze, but I was too busy watching the Bosche to be able to take care of it. I had Mrs. Tanner's helmet on under my cap, but it didn't cover my nose and cheeks.

The Bosche didn't come very near to our part of the formation, but my friend was in a machine that was having a hard time to keep the altitude and once when they (my friend) got about 3000 feet below us the three Bosche machines rushed up and they had a little battle, but no damage was done.

The anti-aircraft or the 'archies' as we call them kept right after us all the time, and at times they broke so close that you could hear the air rush by you with a terrific 'woog-u-f, woog-u-f' and the plane would rock up and down for a short time.

It looked good to see the line come into sight again for all our troubles stopped as soon as we got across and we were soon piling down on the old home ground like a flock of ducks into a lake. My face was covered with ice when I got down and that night my nose, chin and cheeks began to turn black. For about a week I was a sorry looking sight and my nose worried me for a day or so, but it is quite well now.

That night my friend didn't show up at dinner as his machine had to make a forced landing just this side of the lines. The French captain told me to be ready at half-past six in the morning to go out again, as I piled into bed pretty early, but the bombardment going on at the front was unusually heavy and I didn't get much sleep and by morning it was terrific. Got a quick breakfast and at seven was out on the field ready to go, but a fog held us up until 10 o'clock.

About an hour after starting we were just at the line of trenches and there was so much artillery smoke that I couldn't see the ground. Just as we got up there my pilot turned around and said that the engine wasn't working right and that he was going back. The engine seemed to be working all right as far as I could see, but back we went. We should have gone southeast to get back, but he insisted on going south. I tried to convince him to go southeast, but as he didn't speak much English he wouldn't be convinced.

After traveling south for an hour he decided he was lost and picked out a field with a couple of hangars on it to land in. He didn't circle the field to give it the once over, but came right in. Just as he was coming on the field he discovered a road with telephone poles that ran across it and he had to raise to get over it.

By this time he was so low he had to land and so high that he had to drop quick and I thought sure we were going to smash right in the woods. He hit the ground with a smash, a wheel blew up and we plowed into the ground. We didn't get a scratch but I was relieved to get out as we were still carrying our 600 pounds of bombs. We were fortunate to have landed right near a camp of French Chasseurs and they treated us great. We stayed there two days when a machine flew over with a wheel and a mechanic to fix us up.

I suggested to the French lieutenant that he have the mechanic ride back with him as he (the lieutenant) felt that his engine wasn't working right. He thought it was a good idea and so I crawled in with the French sergeant. We no sooner got up in the air than a fog began to roll by and although we were traveling off of my map, I felt we were going too far east, but the pilot didn't think so. We finally had to land to get our directions and although it turned out that I was right, yet we had run out of gas and had to put up for the night.

FIRST TO BOMB

We had landed back of Verdun at the edge of San Sermaize. All the buildings in this city had been burned by the Germans early in the war, in their search for French officers, and about the only building standing was a beautiful chateau and it was in this chateau that we were billeted for the night.

Never slept in quite as gorgeous a room before. The next day I got a good view of how the Germans had burned the city and it can't help but make a person fighting mad to listen to some old French woman tell how they went into her house, piled all the chairs on top of the table, broke a lamp over the pile and set fire to it. They have all built little shacks near the ruins of their former houses and are at least living on the same ground.

Got back to camp the same day and returned home a few days afterwards. Before you get this letter expect to be up and at it again, for when I got back here I learned that they are going to send my friend and two others out from the instruction staff with the first bombing squadron permanently so if you read about it you will know that I am there.

Am going to have a few days vacation and then for the big show again.

FIRST TO BOMB

FIRST TO BOMB

Acknowledgements

Many people aided this project. First, of course, is Francis C. Rath the grandson of Lt. Howard G. Rath. He very generously lent me, a stranger, the diary for an extended period of time so it could be transcribed. That was a great privilege. Michael O'Neal, Managing Editor of *Over the Front* journal, cheerfully shared information from his vast knowledge and has been an enthusiastic supporter of this and other projects. Steven Suddaby provided insights and encouragement, as well as his astonishing bombing database. Novelist, historian, and good friend Jim Littlefield was behind the project from the start and provided a bolt of lightning when it was badly needed. Greg VanWyngarden clarified details and pointed me to sources. Barrett Tillman provided invaluable information regarding the court martial of the Billy Mitchell records. John L. Smith was always eager to supply encouragement as was Louise S. Thoman. Shirley I. Tallon provided endless inspiration insisting it was my duty to transcribe and work with the diary that had so providentially been put in my care. Many thanks to military historian Michael Schellhammer. Other kind and helpful souls include Carl Bobrow and Greg VanWyngarden. Doubtless there are many other individuals who have assisted me in the two years I have been working on this project. To those I apologize and express my thanks. This book would never have become reality without the proofreading, content critiques, rewriting, infinite computer savvy and support of my wife, Sue, who brought it all together and made it happen. It is her book.

FIRST TO BOMB

FIRST TO BOMB

Bibliography

Books

Aircraft Hearings Before the President's Aircraft Board, Washington, DC, Government Printing Office, 1925, vol. 1, p. 1100-1113. https://babel.hathitrust.org/cgi/pt?id=uc1.$b633809&view=1up&seq=8

Arnold, H.H. *Global Mission*, (New York, Harper & Brothers, 1949). Arnold makes a strong case for Air Power.

Boyne, Walter J., *The Influence of Air Power Upon History*, (Gretna, LA: Pelican Publishing Company, 2003).

Codman, Charles, *Contact*, (New York: Little Brown & Company, Boston, 1937). Recommended.

The Forester, Vol. 10, 1907, Lake Forest College, IL., 1907.

Gorrell, Col. Edgar S., *Gorrell's History of the A.E.F. Air Services 1917-1919*. https://www.fold3.com/title/80/gorrells-history-aef-air-service This is the best, by far, source for all things connected with the air war in World War I. As it was written by participants during and immediately after the War, it is rightly considered a primary source. Per NARA: On the 58 microfilm rolls of this publication are reproduced 282 bound volumes of historical narratives, reports, photographs and other records that document administrative, technical and tactical activities of the Air Service in the American Expeditionary Forces. These records – originals, carbon copies, and transcribed copies – are part of Records of the American Expeditionary Forces (World War I), 1917-1923, Record Group 120, publication M990. The history was compiled, not written, by Gorrell or the 119 officers, enlisted men, and civilians engaged in creating the history. The massive compilation by Gorrell was never published and remains the most nearly complete documentation of the history of the AEF Air Service. Vo. E-14 Sq. histories – and N-16 "first army material, histories of the 11th, 20, 96th and 116th sq. operations" and M-12 Lights and Landings No. 24, p. 9, 96th aero bombings squadron: interesting Notes from Sgt. Herbert C. Faust."

The official history of the 96th Squadron was written by Bruce C. Hopper and David H. Young, both pilots in 96th Squadron.

Gorrell – Series J, vol. 4, p. 1-350. "Training. Report on Day Bombardment Training in the AEF." https://www.fold3.com/image/19196048. 350 pages including manual for the 7th AIC bombsight. This would be the basis of the course at Clermont-Ferrand, 7th Aviation Instruction Center, at least in the fall of 1918, but perhaps not in the early spring of 1918. The manual for the 7th AIC bombsight may not be found anywhere else.

Gue, Benjamin F., *Biographies and Portraits of the Progressive Men of Iowa* (Des Moines, Iowa, 1899), Conaway & Shaw Publishers, p. 498-499.

Handbook of Ordnance Data, November 15, 1918 (Washington: Government Printing Office, 1919). https://archive.org/details/handbookordnance00unkngoog.

Harrington, Hugh T., *Destiny's Wings, Four Months in Day Bombardment: The Story of Lt. Hugh S. Thompson, 96th Aero Squadron, U.S. Army Air Service in World War I*, (privately printed, Gainesville, Georgia, 2019). Recommended.

Harvard's Military Record in the World War (Boston: Harvard Alumni Association, 1921).

Hudson, James J., *Hostile Skies, A Combat History of the American Air Service in World War I* (Syracuse, NY: Syracuse University Press, 1968). Recommended.

Inquiry into Operations of the United States Air Services, Hearing before the Select Committee of Inquiry into Operations of the United States Air Service, House of Representatives, Sixty Eighth Congress, on Matters Relating to the Operations of the United

States Air Services. Part 3. Washington, DC, Government Printing Office, 1925.

Martel, Rene, translated by Allen Suddaby, edited by Steven Suddaby, *French Strategic and Tactical Bombardment Forces of World War I,* originally published, in French, in 1939, (Lanham, MD: The Scarecrow Press, 2007). Excellent.

Mitchell, William, *Memoirs of World War I*, (New York: Random House, 1960).

Ticknor, Caroline, ed., *New England Aviators, 1914-1918*, volume 1 (Boston: Houghton, Mifflin Company, 1919).

Thomas, Gerald C., Jr., *The First Team: Thornton D. Hooper and America's First Bombing Squadrons*, (Dallas, TX, The League of World War I Aviation Historians, 1992.) This is an excellent resource.

University of Pennsylvania, *Pennsylvania: A Record of the University's Men in the Great War* (Philadelphia, 1920).

Way, D.H., *Descendants of Robert and Hannah Hickman Way of Chester County*, volume 2, privately printed, 1975.

Waller, Douglas, *A Question of Loyalty: Gen. Billy Mitchell and the Court-Martial that Gripped the Nation*, (New York: Harper Collins Publishers, 2004). Excellent resource for all matters concerning the court-martial and the life of Billy Mitchell.

Who's Who in American Aeronautics, (New York: The Gardner, Moffat Co., Inc., 1922)

Williams College in the World War, published by the President and Trustees of Williams College, 1926, p. 221-3.

Newspapers (in date order)

"Hampton Against Ackley," *Evening Times-Republican* (Marshalltown, Iowa), March 27, 1902, p. 3. [Ackley high school debating club]

"Hardin County Athletes," *Evening Times-Republican* (Marshalltown, Iowa), May 29, 1902, p. 3. [Howard Rath elected vice president of the Hardin County High School Athletic Association]

FIRST TO BOMB

"Ackley Won Debate," *Evening Times-Republican* (Marshalltown, Iowa), February 9, 1903, p. 3.

The Stentor (college newspaper), 1904/5, Lake Forest College, Lake Forest, IL.

Los Angeles Express, September 2, 1910, p. 21. Advertisement, Rath joins Louis Blankenhorn Company forming "Blankenhorn & Rath."

Los Angeles Evening Express, January 25, 1911, p. 20. [Rath elected to L.A. Stock Exchange]

"Farewell Reception," *Los Angeles Express* (Los Angeles, California), May 9, 1914, p. 10.

"Leave on Tour of World," *Los Angeles Express* (Los Angeles, California), May 18, 1914, p. 2.

"F.B. Sayre to Leave for North Tonight," *Los Angeles Express* (Los Angeles, California), May 27, 1914, p. 21.

"In Yokohama," *The Los Angeles Times* (Los Angeles, California), June 7, 1914, p. 36.

"John Rath Dies at Ackley, Age 78," *The Courier* (Waterloo, Iowa), Jun 22, 1914, p. 10.

"Arrive in Yokohama," *Los Angeles Express* (Los Angeles, California), June 22, 1914, p. 14.

"Arrived in Iowa With Only Cents, John Rath Died Worth Half Million Dollars," *The Courier* (Waterloo, Iowa), June 23, 1914, p. 12.

"New European route Preferred by Many Los Angeles People," *Los Angeles Express* (Los Angeles, California), July 8, 1914, p. 12.

The Los Angeles Times (Los Angeles, California), advertisement: Blakenhorn and Rath dissolves and Howard G. Rath Company starts, September 15, 1916, p. 20.

The Los Angeles Times (Los Angeles, California), Howard G. Rath Company advertisement, January 10, 1917, p. 21.

The Los Angeles Times (Los Angeles, California), Howard G. Rath Company advertisement, January 21, 1917, p. 76.

"'Judy' Rath Into It," *University Club Bulletin*, Los Angeles, California, Vol. 2, No. 1, April 1, 1918. p. 3.

"Bombing the Boches, Sport for Angeleno, Former Stock Exchange Man Tells of Aerial Thrills – Rheims is Like Volcanoes on

Moon," *The Los Angeles Times* (Los Angeles, California), April 28, 1918, p. 21.

"Writes of Air Battle Thrills, Former Stock Exchange Member Describes Hair-Raiser, He and Pilot Sole Survivors of Fight with Huns, Engine Fails on Bombing Trip, Escape by Hairbreadth," *The Los Angeles Times* (Los Angeles, California), October 27, 1918, p. 25.

"Wins Cross For Bravery in Air," *Oakland Tribune* (Oakland, California), Nov. 5, 1918, p. 12.

"Weeks is Puzzled, Secretary of War Ponders Over General Mitchell's Reply to Charges: Wilbur Gains Some Support," *The Pomona Progress* (Pomona, California), February 9, 1925, p. 1.

"Weeks To Study Charges of Mitchell," *Hanford Sentinel* (Hanford, California), February 9, 1925, p. 1.

"Reappointment of Mitchell Deferred Until Probe is Ended, Weeks Desires to Have Full Data Before Acting – Former Air Officers Offer to Tell of General's 'War Conduct,'" *Evening Star* (Washington, DC), February 9, 1925, p. 1.

"Recall Mitchell on Bomb Tests," *Evening Star* (Washington, DC), February 11, 1925, p. 1.

"L.A. Bombing Expert Ordered to Air Probe, Howard G. Rath Will Testify Before Coolidge Board," *Los Angeles Evening Express*, (Los Angeles, California), September 30, 1925.

"L.A. Man to Testify in Aircraft Probe," *The Bakersfield Californian* (Bakersfield, California), September 30, 1925, p. 1.

"Rath Back from Air Service Inquiry," *Los Angeles Evening Express* (Los Angeles, California), October 24, 1925, p. 24.

"Rath Gives Air Board Testimony, Los Angeles Bond Dealer Assails Mitchell; Is Back From Eastern Trip," *The Los Angeles Times* (Los Angeles, California), October 28, 1925, p. 14.

"Los Angeles Man Called as Witness," *The Los Angeles Times* (Los Angeles, California), November 21, 1925, p. 2.

Bennett, James O'Donnell, "Reid Enraged by Hero Witness, Hissed by Crowd, General of Court Defends Critic of Mitchell," *Chicago Tribune* (Chicago, IL), December 10, 1925, p. 7.

"Patrick Upholds Mitchell in Part," *Evening Star* (Washington, DC), December 10, 1925, p. 1, 2.

"Assails Mitchell on Deaths in War," *New York Times* (New York, NY), December 10, 1925, p. 10.

"Attack on Testimony of Flyer Fails," *The San Francisco Examiner* (San Francisco, CA), December 10, 1925, p. 3.

"Mitchell Assailed for Orders Issued During World War," *The Atlanta Constitution* (Atlanta, GA), December 10, 1925, p. 1.

"Mitchell is Criticized, Army Men Attack His Judgment In Order During St. Mihiel Battle," *The Miami Herald* (Miami, FL), December 10, 1925, p. 3.

"'Almost Treason' Charge is Denied in Mitchell Case," *Evening Star* (Washington, DC), December 12, 1925, p. 1, 4.

Urich Herald-Montrose Tidings (Urich, Missouri), December 17, 1925, p. 6. This untitled article appeared in many newspapers across the country.

"Introducing Deming AAF Officers, Captain Howard G. Rath," *The Deming Headlight* (Deming, New Mexico), February 26, 1943, p. 1.

"Last Rites Set for Broker Howard Rath," *The Los Angeles Times* (Los Angeles, California), May 3, 1973, p. 47.

Rath, Helen Cowell, obituary. *The Los Angeles Times* (Los Angeles, California), Nov. 5, 1997, p. 90.

Rath, Howard G., Jr., obituary. *The Los Angeles Times* (Los Angeles, California), February 26, 2017, p. B8.

Magazine and Journal Articles

Harrington, Hugh T., "We Were Told What to Write, a Closer Look at Gorrell's History," *Over the Front*, The League of World War I Aviation Historians, vol. 36, no. 1, Spring, 2021, p. 4-17.

Hickam, the Cavalry Convert, Air Force Magazine, November/December 2016, p. 80. https://www.airforcemag.com/PDF/MagazineArchive/Magazine%20Documents/2016/November%202016/1116namesakes.pdf. Accessed October 22, 2020.

Hopper, Bruce C., "American Day Bombardment in World War I," *Air Power Historian*, April 1957 p. 87-98. Recommended.

Johnson, Alfred W., Captain, USN, "The Naval Bombing Experiments: An Account of the Bombing," *Naval History and Heritage Command*, retrieved April 29, 2020.

Johnson, Alfred W., Vice Admiral, USN, "The Naval Bombing Experiments Off the Virginia Capes – June and July 1921," *Naval History and Heritage Command*, 1959, retrieved April 29, 2020.

Leiser, Edward L., compiler, "Red Devil in a Breguet, David H. Young, 96th Aero Squadron," *Cross & Cockade*, vol. 12, no. 2, Summer 1971, p. 155-168. Transcription of taped interviews with 96th Squadron pilot David H. Young. Recommended.

Miller, Roger G., "Billy Mitchell, the 3rd Attack Group and the Laredo Project," *Air Power History*, Summer 2007, vol. 54, number 2, p. 4-15.

Mitchell, William, "Aeronautical Era," *Saturday Evening Post*, December 20, 1924, reprinted in *Inquiry into Operations of the United States Air Services, Hearing Before the Select Committee of Inquiry, House of Representatives, Sixty Eighth Congress, on Matters Relating to the Operation of the United States Air Services*, Washington, DC, Government Printing Office, 1925, p. 2032-2039.

Rath, Howard G., "Needed – Mine Sweepers of the Air," *U.S. Air Service,* Army and Navy Air Service Association, vol. 1, no. 3, April 1919, p. 18-22

Suddaby, Steven, "Aerial bombing in World War I," *Over the Front*, League of WWI Aviation Historians, Volume 35, Number 4, Winter, 2020, p. 316-331. Recommended.

Other

Arlington National Cemetery. http://www.arlingtoncemetery.net/hmhickam.htm. Accessed October 22, 2020. Colonel Horace M. Hickam.

Dennett, Tyler, Passport Application, April 30, 1914. NARA, Washington D.C.: Roll# 210; Certificates 28708-2966, 04 May 1914-12 May 1914. Ancestry.com US Passport Applications, 1795-1925 [database online]. Lehi, UT, USA.

FIRST TO BOMB

Hopper, Bruce C., Captain, 96th Aero Squadron, A.E.F., *When the Air Was Young, American Day Bombardment, A.E.F France, 1917-18*, typescript dated January-February, 1919. Includes Preface by Bruce C. Hopper, Historian, USSTAF, June 21, 1944. Four-part narrative includes chapters on the history of American Day Bombardment, a Manual of Instructions, History of the 96th Aero Squadron, by Captain Bruce C. Hopper and an Appendix of the Operations Log of the 96th Aero Squadron. Unpublished. Located in Library of Congress.

Mitchell, William, *Court Martial Trial Transcript* is located at the NARA facility at College Park, MD. The transcript is 3781 pages. Also included are literally thousands of court documents, exhibits, correspondence, etc.

Mitchell, William, *Court Martial Trial Transcript*. NARA M1140. Secret and Confidential Correspondence of the Office of the Chief of Naval Operations and the Office of the Secretary of the Navy, 1919-1927. The transcript is available at NARA online, however, it is very difficult to navigate. It can also be found online at fold3.com and is somewhat easier to navigate. The Testimony of Howard G. Rath took place on December 9, 1925. It can be found on pages 2935-3007. See www.fold3.com/image/296694606.

Officer's Record Book. A.G. Printing Dept., G.H.Q., A.E.F., 1915. This 3.5" x 5.5" booklet of 8 pages was supplied to officers to enable them to record their service, courses of instruction, promotions, awards and "Remarks" by superior officers on the performance of the booklet holder. Rath's record book was found with his Diary, newspaper clippings and other small items.

Passenger List. *Olympic*, departed Liverpool August 22, 1914, arrived New York, August 30, 1914. NARA 1914, arrival: New York, New York; Microfilm Serial: T715, 1897-1957; Microfilm Roll: Roll 2365: Line: 15; Page Number: 109. Ancestry.com.

Passenger List. *Lusitania*, departed Liverpool, December 16, 1914, arrived New York, December 24, 1914. NARA 1914; Arrival: New York, New York; Microfilm Serial: T715, 1897-1957; Microfilm Roll 2391, line 23, page number 91. Ancestry.com.

Passenger List. *Saxonia*, Cunard Line, Liverpool arrival October 9, 1917. Howard Grant Rath, Broker. National Archives of the UK;

Kew, Surrey, England; Board of Trade: Commercial and Statistical Department and successors: Inwards Passenger List: Class: BT26; Piece: 638. Ancestry.com. UK and Ireland, Incoming Passenger Lists, 1878-1960 [database online]. Provo, UT, USA: Ancestry.com Operations Inc., 2008.

Passenger List. *Stockholm*, departing Brest, France, February 2, 1919, arriving New York February 12, 1919. NARA Lists of Incoming Passengers, 1917-1938. Record group 92.

Passenger List. *Olympic*, White Star Line, departing Southampton, August 23, 1922. National Archives, London, England. www.nationalarchives.gov.uk.

Passenger List. *Olympic*, arriving New York from Southampton, August 29, 1922. 1st class. Rath address: 2966 Wilshire Bldg, Los Angeles.

Passenger List. SS *Rotterdam*, Holland American Line, arrive New York September 3, 1926 from Boulogne, France. NARA. Year: 1926; Arrival: New York; Microfilm Serial: T715, 1897-1957; Microfilm roll 3917; Line 3, page 94. Ancestry.com.

Passenger List. Trans World Airlines, Departure July 31, 1954, New York to Frankfurt, Germany. NARA Passenger and Crew Lists of Vessels and Airplanes Departing from New York, 7/1/1948-12/31/1956; NAI Number: 3335533; Record Group title: Records of the Immigration and Naturalization Service, 1787-2004; Record Group number: 85; Series Number: A4169; NARA Roll Number: 274. Ancestry.com.

Passenger List. *Nieuw Amsterdam*, arriving New York, October 16, 1954. From Southampton, departure October 9, 1954. First Class. NARA 1954; Arrival: New York; Microfilm Serial: T715, 1897-1957; Microfilm roll 8521; Line: 4; Page 282. Ancestry.com.

Passenger List. Ship MS *Gripsholm*, Swedish Line. Arriving New York, April 29, 1961, departed Lisbon, Portugal April 22, 1961. NARA NAI number: 2990227; Record Group Title: Records of the Immigration and Naturalization Service, 1787-2004; record Group Number: 85; Series A4115; roll number 649. Ancestry.com.

Rath, Howard G., "Origin and Development of American Aerial Day Bombing During World War I," unpublished typescript, undated, circa 1940s. It is in the custody of the Rath family.

Rath, Howard G., Personal Diary, the daily diary on 256 pocket sized pages. Original in possession of the family.

Rath, Howard G., Passport Application, April 30, 1914. NARA; Washington D.C.; Roll #:210; Certificates: 28708-2966, 04 May 1914-12 May 1914. Ancestry.com US Passport Applications, 1795-1925 [database online]. Lehi, UT, USA

Rath, Howard Grant, Passport Application, September 11, 1917. NARA; Washington D.C.; Roll#:401; Certificates: 64901-65200, 08 Sep 1917-10 Sep 1917. Ancestry.com. U.S. Passport Applications, 1795-1925 [database online]. Lehi, UT, USA: Ancestry.com Operations, Inc., 2007.

Rath, Howard G., Passport Application, April 5, 1922. NARA; Washington D.C.; Roll #: 1888; Certificates: 139350-139725, 05 Apr 1922-05 Apr 1922. Ancestry.com. U.S. Passport Applications, 1795-1925 [database online]. Lehi, UT, USA: Ancestry.com Operations, Inc., 2007.

Rath, Johannes, Baptism record, 27 Nov. 1840. [John Rath, father of Howard G. Rath] Ancestry.com Wurttemberg, Germany, Lutheran Baptisms, Marriages, and Burials, 1500-1985 [database online]. Provo, UT, USA: Ancestry.com Operations, Inc., 2016. Original data: Lutherische Kirchenbucher, 1500-1984. Various Sources.

Rath, Howard G. and Helen C. Kelly, Marriage Record, March 21, 1927, Los Angeles, California. California Department of Public Health, courtesy of www.vitalsearch-worldwide.com Digital images.

Internet Resources

Suddaby, Steven, *Suddaby Western Front Bombing Database*, www.overthefront.com, accessed May 11, 2020. This massive and comprehensive database covering every bombing mission of the War is an outstanding resource.

Index

[unknown last name]

Capt. C 40
Charley the Deck Steward 12
Elizabeth 68
George (Lt.) 114, 122
H.C. 82
Han [?] 83
Jim 90
Joan 25
John 116
Mrs. C. 70
Simeon 114
Suzanne 74

A

Alexander, Arthur "Alex" 77, 79, 89, 90, 116
Allsopp, Clifford 112
Alsford, Lt. 86, 87
Anderson 13
  Charles Patrick "Pat" ... 47, 53, 55, 63, 65, 84, 86, 87, 88, 94, 96, 98, 99, 103, 104
Anspach, Ralph 79, 94, 106
Arbuckle, Fatty 87
Atkinson, Col. 119

B

Bates, Edmond 90, 94
Bell, Major General 8
Bernhardt
  Maurice 36
  Sarah 36
Beverley, Belmont F. 84, 93, 99, 110, 113, 115, 121
Blackshear, Dr. 22, 23
Blakeman
  Capt. 24
  Frederick T. 15, 24
  Lt. 30, 35, 74
Bleeker, Lyman C. 94
Bourjolly, Lt. de 34
Bowen, Thomas S. "Major B" .. 112, 113, 114, 115, 116, 122
Boyd, Capt. 22
Brainard, Spencer. 23, 24, 27, 28, 30, 31, 106
Breen, [unknown first name] 71
Brown, Harry M. "Major B" .. 37, 39, 40, 44, 50, 54, 55, 61, 63, 64, 65, 66, 67, 68, 125
Browning, Robert C. 66
Buckler, Lt. 25

C

Carter, Arthur S. 105
Cawston, Arthur H. 48, 94, 99, 106

Chapin, Roger F. 99, 105
Chaplin, Charley............................... 8
Clapp, Roger.......... 30, 35, 47, 48, 49, 63
Codman, Charles R..... 43, 47, 67, 69, 73, 79, 80, 81, 84, 86, 90, 91, 94, 98, 103, 117
Cody [unsure whether first or last name] 40
Comegys, Edward T. 105
Cook, Lt.. 121
Cooper, Johnny............................. 74, 90
Coryell, Ralph I. "Cory" 91, 113, 115, 118, 121
Crimmins [?] or Krummins [?]. 111, 118, 121
Cronin, Edward M. 94

D

Driggs, Mr. 119
Duke, James E. Jr. 66
Dunn, Robert J..................................... 63
Dunsworth, James L... 66, 69, 70, 72, 73, 89, 90, 91, 95, 104, 106, 110, 111, 112, 113, 114

E

Elliott
   R.P. "Bob" 102, 122
   Walter Keith 8, 13
Ellis, Arthur C. 37, 39, 102
Estes, Capt... 119
Evans, Elisha E. "Chick" 40, 41, 43, 50, 60, 68, 69, 74, 75, 82, 84, 86, 87, 88, 89, 90, 94, 97, 103, 104, 105, 107, 108, 111, 114, 115, 116, 119

F

Fairbanks, Douglas 8
Fairlamb, Miss................................... 44
Farnsworth, Thomas H. 43, 47, 50, 60, 75, 77, 84, 85, 86, 87, 90, 96, 97, 98, 104
Fitzgerald.. 22

Foch, Ferdinand............................... 118
Forshay, Harold J. 109
Foulois, Benjamin D. 66

G

Garrett 22, 23
Gatton, Cyrus J................................. 117
Gaylord, Bradley J...... 60, 75, 76, 79, 80, 84, 87, 88, 90, 93, 96, 97, 98, 99, 100, 102, 114, 118, 119
Gosling, Eleanor............................... 107
Graf, Bob... 22
Green, Robly 24, 27, 28, 30, 31, 35
Greene, Lt................................. 22, 23, 24
Grey, Alexander 50, 61, 63, 64
Gros, MD, Edmund L. 15
Gundelach, Andre H. "Gundy"..... 22, 24, 25, 41, 47, 48, 53, 59, 63, 69, 70, 72, 73, 74, 77, 79, 80, 82, 83, 84, 85, 86, 87, 88, 89, 90, 91, 93, 94, 103
Guthrie, Ramon 104, 105

H

Hammond.. 23
Hanby (or possibly Haley) 20
Handley [?], Lt. 40
Harris, J.J.. 100
Hart, Bill 87
Harter, Lester S. 105
Heater, Charles L. 116
Heninger [?] 41
Hexter, Avrome...... 89, 90, 93, 104, 107, 108, 111
Hooper, Thornton D. "Hoop" or "Nap"... .47, 49, 50, 59, 60, 61, 62, 63, 65, 67, 69, 70, 71, 72, 74, 75, 76, 77, 84, 87, 89, 90, 100, 104, 105, 117
Hopkins, Stephen 96, 97
Hopper, Bruce C... 98, 99, 112, 113, 120, 121, 122
Howard
   Frank J. 109
   unidentified 24
Hower, Virgil H. "Val"............ 69, 73, 74

J

Janis, Elsie 47

K

Kelly
   Arthur H. 89, 99, 112, 113, 116
   Capt. 22
Kelsey, Florence 7
Kinsolving, Charles 29
Knisely, George 11
Krummins .... *See* Crimmins
Kyle, George 29, 30, 35

L

LaForge, Edward C. 13
Laird, Clair B. 99, 105
Lakin 122
Lane, Lewis Palmer 11
Lee
   [unknown first name] 23
   William A. "Doc" 104, 105, 106, 107, 115
Lehutty, Marion 41
Leroy, C.H. 66, 69, 70, 104, 107, 108, 116, 117
Lewis, Henry C. 55, 66
Liggett, Hunter 118
Lindsay, E.W. 121
Lucas, Natalie 107
Lufbery, Raoul 44
Lunt, Samuel M. 75, 76, 82, 84, 87, 88, 93, 99, 100, 102

M

MacArthur, Miss 116
MacArthur/McArthur, Miss Gladys ...24, 25
MacDonald, Durwood L. 55, 66, 67
Marshall, Frank 41
McBride, Jimmy 40, 43
McChesney, Harold A. 62, 66, 67

McDowell, Stewart A. 96, 103
McLennan, John Charles Earle "J.C.E."
.... 80, 89, 90
McMahan, Miss 70
McNeil, Dr. 27
Mellen, Joseph M. 55, 63, 66, 67
Metzger, Arthur R. 22, 23, 41
Milling, Thomas DeWitt 98, 111, 120
Mitchell, William "Billy" .... .65, 95, 98, 111, 115, 121, 122, 124, 125
Morse, Charles F. Jr. 11
Mounds 49

N

Narden [?], Major 40
Newbury, Frank J. 99, 115
Newell, Hester "Peter" 10, 11, 13
Normand, Mable 87
Normid [?] (the cook) 90
Norris, Lt. 109
Norton 23, 25
   Tom 41, 43

O

O'Donnell, Paul 109
O'Toole, James A. 74, 79, 91
Oatis, Vincent P. 104, 105

P

Pancoast, Henry L. 115
Parrott, E.A. 100, 101, 109
Patton, George C. 103
Payne, Karl C. 99, 101, 102
Petit, Capt. 28
Petre, Capt. 115
Pickford, Mary 8
Pike, John "Bunny" 8, 11, 12, 13, 24, 35, 48, 49, 71, 113
Pressler, Warren S. 60, 70, 79, 85, 91, 106
Prince [?], Lt. 82
Purviance, Edna 8

R

Rader, Ira. A. ... 39, 44, 62, 104, 106, 107
Ratterman, George A. "Rat" 63, 65, 66
Reagan, William N. (Bill) .. 9, 11, 13, 24, 35, 48, 49, 71
Reed
   Capt. 40, 43
   Thomas R. 121, 122
Reel, Capt. 20, 23
Richthofen, Manfred (Baron) von 82
Rickenbacker, Eddie 47, 119
Ring, Lt. 115
Roberts 23
Robertson, Philip "Bobby" 25, 27, 28, 35, 41
Rogers, Newton C. 96, 103
Root, Ralph R. 105, 117
Rory 87

S

Sellers, Cecil G. "Swede" 23, 44, 50, 60, 66, 68, 69, 75, 76, 77, 84, 87, 89, 90, 99, 100, 101, 102, 107, 116, 118, 119
Smith
   Herbert D. "Smithy" 41, 47, 60, 66, 67
   Leigh Hackley 9, 13
Soulin [?], Madame 40
Stearns, William S. 48
Stephens
   C. T. 100
   Robert G. "Steve" 47, 68, 74, 106, 107, 116, 118, 121
Stephenson, McRea 105
Strafford, Lady 36
Strauch, Harry H. 100, 105
Strawn, Kenneth P. 96, 103, 117
Strobbing [?], Lt. 39
Strong, Alfred R. 55, 66
Stuart, William A. 91, 94, 103, 117
Summersett, James A. 60, 63, 65, 66, 68, 73, 75, 76, 93, 94, 115, 121, 122

T

Taylor
   James J. 8, 11
   Raymond C. 103, 117
Thibaud [?], Lt. 44
Thomas, George C. 22, 23, 37, 38, 40, 41, 61, 62, 66, 68, 69, 77, 82, 83, 114, 116, 118, 120, 122
Thompson
   Hugh S. 47, 51, 55, 70, 87, 88, 89, 94, 99, 103, 117
   Robert E. ... 63, 70, 77, 85, 87, 89, 91, 96, 97
Tichener, Claxton H. 55
Tom [?], Miss 36
Tucker, Rowan H. 55, 66
Turnbull, Lewis F. 94, 113, 121
Tyler, John C. 105

V

Valentine 22
Van Sickle, Steve 40, 41, 43, 44
Verity, Hun 7
Vuillemin, Commandant 28

W

Walden, James P. 112
Ward
   Capt. 50, 61, 63
   Lawrence 112
Warner, Donald D. 89, 90, 116, 121
Way, Pennington H. "Pinky" .. 16, 20, 23, 24, 25, 27, 28, 29, 30, 35, 37, 41, 43, 48, 49, 59, 63, 67, 68, 69, 70, 71, 72, 74, 77, 79, 82, 83, 84, 85, 86, 87, 88, 90, 91, 93, 94, 95, 98, 103, 107, 128
Weeks, John W. 124
Wilbur, Curtis D. 124, 125
Wilhelm II, Kaiser 24, 120
Williams, Bertram 96

Wilson
  Lillian 74
  Miss 36
  Woodrow (president) .. 115, 116, 117, 118
Woolf, William B. "Wolf" 12, 13
Woolley 23
Wright, General 13

Y

Young
  Cecil P 79, 80, 84, 87, 90, 106
  David H 51, 75, 76, 82, 84, 85, 86, 87, 88, 93, 99, 101, 110, 113, 114, 115, 121

www.ingramcontent.com/pod-product-compliance
Lightning Source LLC
Chambersburg PA
CBHW031252290426
44109CB00012B/543